A Guide to Understanding the Fundamental Principles of Environmental Management

It Ain't Magic: Everything Goes Somewhere

A Guide to Understanding the Fundamental Principles of Environmental Management

It Ain't Magic: Everything Goes Somewhere

Andy Manale and Skip Hyberg

Published by   IWA Publishing
                      Republic – Export Building, 1st Floor
                      2 Clove Crescent
                      London E14 2BE, UK
                      Telephone: +44 (0)20 7654 5500
                      Fax: +44 (0)20 7654 5555
                      Email: publications@iwap.co.uk
                      Web: www.iwapublishing.com

First published 2021
© 2021 IWA Publishing

Disclaimer
The information provided and the opinions given in this publication are not necessarily those of IWA and should not be acted upon without independent consideration and professional advice. IWA and the Editors and Authors will not accept responsibility for any loss or damage suffered by any person acting or refraining from acting upon any material contained in this publication.

British Library Cataloguing in Publication Data
A CIP catalogue record for this book is available from the British Library

ISBN: 9781789060980 (paperback)
ISBN: 9781789060997 (eBook)

Contents

Foreword

Technological innovation, which has given humans dominance over land and water and made once scarce resources readily available, is a two-edged sword: it has freed much of humanity from the shackles of food and resource scarcity, but it does so at the cost of environmental degradation. The latter comes, ironically, from excesses of what makes life possible.

The research that I and my fellow scientists conduct documents the adverse impacts of ill-management of natural resources and illuminates solutions to the dilemma. Nevertheless, if progress towards sustainability is to be achieved, the science must be translated into action. In other words, managing and protecting the environment means making decisions on how resources are used. Not all decisions are alike. We naturally want the action to solve the problems we face, but the likelihood that it does so depends substantially upon how well the problem has been defined and the appropriateness of the response. There is no set script to guide actions for the manager to follow. Scientific knowledge is ever evolving in the context of ever-changing human demands on natural resources made ever more complicated within a changing climate. The successful manager must know how to construct responses to these difficult challenges on sound scientific principles. He or she must know or anticipate, as if instinctively, where the sword will slice, and how to seek appropriate scientific guidance.

Andrew Manale and Skip Hyberg provide an excellent introduction to today's enduring and existential environmental problems, as well as the key scientific principles underlying them. The latter are, regretfully, too often neglected. The authors have extensive experience in the trenches of programmatic and policy battles over whether or not and how to achieve environmental sustainability.

Their efforts have not gone without considerable success. I know this is to be so. I have known and worked with them for many years in many contexts. Their book, targeting the introductory student, the lay person looking for insight into the sticky problems of today, and the policy and manager professional needing a core grounding in the science, provides a framework for identifying what can work towards achieving the goal of an environment in balance and how to measure progress. It should be read broadly.

Rattan Lal

Distinguished University Professor of Soil Science, SENR

Director, CFAES Dr. Rattan Lal Carbon Management and
Sequestration Center

IICA Chair in Soil Science and Goodwill Ambassador
for Sustainable Development Issues

Adjunct Professor at the University of Iceland and the
Indian Agricultural Research Institute (IARI)

Past President, International Union of Soil Science
December 8th 2020

Preface

*Possibly, in our intuitive perceptions, which may be truer than our science
and less impeded by words than our philosophies, we realize the indivisibility
of the earth – its soil, mountains, rivers, forests, climate, plants, and animals,
and respect it collectively not only as a useful servant but as a living being.*

— Aldo Leopold (from Thomas Tanner, ed., Aldo Leopold,: the Man
and His Legacy, Soil and Water Conservation Society, 2012)

If you are a beginning student in environmental sciences (at college or postgraduate level) or a beginning practitioner of environmental management, you should read this book. The intended audience also includes nonscientists with a keen interest in the **environment** who will acquire an introduction to the seemingly intractable, so-called sticky environmental and **natural resource** issues vexing modern society. Using plain English, we explain the core scientific principles underlying environmental management – everything goes somewhere and **gravity** is omnipresent–, presuming only a rudimentary background in the basic sciences. We build upon the readers' innate knowledge of these principles to provide a framework for understanding how elemental components of the natural world interlock. Stressing the importance of observation (and measurement as necessary), we provide tools for identifying options in managing problems and, by building upon what they know, bolster readers' confidence in applying these.

© IWA Publishing 2021. A Guide to Understanding the Fundamental Principles of Environmental
Management. It Ain't Magic: Everything Goes Somewhere
Authors: Andy Manale and Skip Hyberg
doi: 10.2166/9781789060997_xi

Readers will learn to ask questions that help unravel the complexity of environmental and **resource management**. You will learn to see how seemingly disparate headline-grabbing issues, such as ever more severe flooding, persistent water quality problems and even **climate change**, interconnect. We use basic insights, "easy rules," that use common sense and direct observations/experiments (empirical knowledge), to explain the core scientific principles underlying environmental management and how these relate to the basic functioning (the balancing of environmental cycles) of our natural world. The book draws upon case studies ("stories") derived primarily from temperate **ecosystems** in and around agricultural **systems** and the modified, i.e., human-altered environment. The examples apply broadly not just to the United States but also to other parts of the world with developed agroecosystems. Experiments engage the reader in observational exercises that use their personal experience to reinforce and extend their understanding of the basic principles.

The framework is not just an academic exercise. It is a tool that can be employed to avoid environmental missteps and make **conservation** efforts more effective. We give examples illustrating the critical role that observation and monitoring play in assessing the effectiveness of a plan and the power of environmental accounting. We will help the reader see how soils provide critical environmental services that support the quality of air and water and interact with every component in our **biosphere**.

The first chapters provide an introduction to key scientific concepts and then relate this science to issues of environmental and natural resource management. Later chapters guide the reader along in seeing how human actions affect natural resource quality and why it matters. The case studies illustrate how poorly conceived management strategies to address one problem can have unintended effects that create larger problems. The case studies show the consequences of not examining the full path of materials as they move through the environment and not making a comprehensive accounting of where they go. Poor decisions, with few exceptions, lead to poor outcomes. The examples give the reader the opportunity to apply the methods and tools in asking what went wrong and what could make the situation right. We conclude with a discussion of when cooperative action is necessary and the role of public **policy**.

With the completion of this book, the reader will be able to recognize how major environmental problems relate to each other and to identify shortcomings of practices and policies for managing them. Most importantly the reader will recognize that to make even small strides toward fixing today's major environmental and natural resource issues requires that the basic principles be incorporated into any strategy.

This book is not is a **chemistry**, **biology**, **ecology**, or **physics** text that grounds the beginning or advanced student in the core knowledge of the respective science. Rather, this book equips you with a basic understanding of core concepts in environmental management and a sensitivity to the complexity of persistent environmental and natural resource problems, challenging you to delve deeper by providing recommendations for additional reading and more detailed answers. In doing so, we draw upon many scientific disciplines, including the social sciences and, from them, distill the essential lessons for the introductory student or informed layperson with an interest in resource management. Although the messages are based upon numerous laws of science, as noted above, we leave more in-depth instruction in these subject areas to others. We demonstrate how drawing upon a few core principles helps you identify the areas of expertise you need to develop for your career, to interact with professionals with the expertise in these sciences, to develop appropriate strategies, and to ask critical questions regarding proposed alternative management options. With these skills, you will understand better the processes affecting climate change, water quality, **floods** and **droughts**, **soil** health, habitat loss, and water quality degradation and how they interrelate. By the concluding chapter, it should be clear to you why and how the above problems act as systems problems that require systems solutions.

About the Authors

The authors have ample real-world experience in probing the sticky, difficult environmental and natural resource issues confounding decision-makers today. They have worked on these issues at every geographic scale, explaining and providing advice to a wide expanse of interested parties, from local community groups to global international governmental and nongovernmental organizations, from farmers and high school and college students to members of Congress and political leaders domestic and abroad. They draw upon their backgrounds in the biological, chemical, and social sciences.

Andrew Manale, M.S. M.P.P., is a retired public policy analyst, biochemist, and researcher who spent many years in the policy office of the United States Environmental Protection Agency and at California's Air Resources Board. He is also a Fellow of the German Marshall Fund and John J. McCloy Foundation, and a LEGIS Congressional Fellow.

Bengt 'Skip' Hyberg, PhD, is a retired economist and scientist from the United States Department of Agriculture's (USDA) Farm Service Agency. During a 32-year career with USDA, he served as a senior advisor to USDA's Chief Scientist and as an Exchange Officer with the Australian Bureau of Agricultural and Resource Economics. He is the recipient of numerous awards for his work, including the Economist of the Year award, the Fred Woods award for sustained excellence in policy leadership and the John E. Lee Award from the USDA Economist Group, the Honors Silver Metal from the Environmental Protection Agency, and the Wetland Conservation Achievement Award from the conservation organization, Ducks Unlimited.

Acknowledgements

We wish to acknowledge the following people whose support has been critical to this project: Jonathan Beebe, who braved the first draft and who helped prepare the citations and glossary; our friend David Hewitt who helped us launch this project and prepared a rough draft for Chapter 2; Peter Kuch, Marc Ribaudo, Mike Linsenbigler, and Jim Johnson, whose reading and comments on a later draft greatly improved the products; Jerry Hatfield for his time and kindness in sharing his experience and knowledge; Stacy Ritcher and Charlie Schafer from the Agricultural Drainage Management Coalition; Andrew Sharpley, our colleague on the Science and Policy Committee of the Soil and Water Conservation Society; Rob and June Wolcott with whom we have worked so many years; and the numerous persons and organizations who have permitted us to use educational graphics in this text. We thank Dr. Rattan Lal for agreeing to read and then to write the foreword to the book. Finally, we wish to acknowledge Professor Bill Ellis, whose support and provision of his student Jonathan Beebe helped make the book possible.

This book is dedicated to our wives, Shizumi and Lauren, who have enriched our lives, nourished and comforted us, and made us better persons. They suffered through our preoccupation and distraction with this book for many months, and endured our rambling and lecturing. This book would not have been possible without their support.

Part I
The Basics or How Stuff Happens

Chapter 1
Introduction

Conservation is a cause that has no end. There is no point at which we will say our work is finished.

— Rachel Carson

All of us experience at times the occasion where we read something that elicits the response: 'That's obvious, I knew that.' And we did. This same deep-down understanding of something to which you may have never been exposed but about which you intuitively know can help fathom environmental issues and judge proposals that address them. We are all scientists in the early years of our lives. As infants we conduct experiments to gauge the workings of our **environment**. To the mother's or father's lament, the child takes the cup and spills the contents on the ground. We learn that things fall when there is no opposing force, i.e. the hand, preventing it from doing so. Do you not remember tossing the contents of a cup into the air and laughing as it projects for a short period of time upward until it succumbs to some force that brings it back down to earth?

We as infants and young adults learn important truths or concepts in general basic ways. What we learn about how our physical world functions form our initial view of the world, that is, our understanding of what happens when an action is taken. Water flowing out of a glass always falls down; it does not go up or float in the air. Our developing brains begin to link an action, with a reaction, a cause with an effect. We know nothing about gravity, at this point in our lives, yet we begin to know what it does to objects. In fact, we can observe a baby drop objects

© IWA Publishing 2021. A Guide to Understanding the Fundamental Principles of Environmental Management. It Ain't Magic: Everything Goes Somewhere
Authors: Andy Manale and Skip Hyberg
doi: 10.2166/9781789060997_0003

intentionally, seemingly to see what will happen. Is the infant experimenting to find exceptions in what is developing as a rule in her perception of her surroundings and how they function? Perhaps this experimentation explains children's love of helium-filled balloons. Or is her laughing because the action has indeed become a game, one where she begins to know the answer and anticipates subsequent actions as her parents respond to the bouncing or splashing outcome? Maybe both. In time, we no longer need to continue this experiment, we intuitively understand this scientific principle. Later, when it is explained to us in school, we grasp the concept immediately. It makes sense since the principle conforms to our experiential understanding of how the world works. Because we already understand the concept, we use this as a central principle that allows us to examine how changes ripple through our environment.

Childhood experimentation also taught us a second central organizing principle for how the world works and, of relevance here, something basic about environmental management – everything goes somewhere. Adults play peek-a-boo with infants. This teaches the concept of object permanence. You learned your parents did not **disapparate** (as Harry Potter would) when they left the room and you know that water does not disappear when it runs off the driveway or soaks into your lawn. This rule applies to all things; this book will help you extend this concept to the core principle – all **mass** is conserved – it is neither created nor destroyed. Intwined in this concept is the fact that although mass is conserved, materials can change physical and or chemical form. You know that even though liquid water can freeze into ice or evaporate, the water is still there. At the more complex molecular level – chemicals change from one form to another. Yet the atoms that comprise the original **molecules** still exist in the same quantity after each change. We will demonstrate to you why this is important in environmental management for you to know where things go. We help you follow where things go even as they change form.

The understanding that we develop at an early age addresses our own particular everyday needs. By rediscovering these tools, we enhance them. You begin to ask questions left unanswered or unexplored earlier and learn how to build new databases necessary for an ever more technical and complicated world. The object moves out of sight. What then? We will explore where it goes and how you will know where to look.

Because you already have a sense regarding how these two principles work, you possess a foundation from which to analyze critical environmental issues. However, they do not suffice for other, more complex ones. The book will add to these two principles a third – natural processes maintain a relatively constant ratio between **carbon**, **nitrogen**, and **phosphorus** in the biosphere and does this by keeping a balance among the atmospheric, oceanic, terrestrial, **soil**, and biologic pools for each **element**. This process operates at different time frames for different elements and media, so disruptions can reverberate for long periods.

With the principles identified and explained, we add a few tools that you already know how to use. They are these: observe, measure, record, ask, and challenge. Through most of this book we present experiments that can be completed simply by the exercise of watching and seeing what happens. Often, it will be demonstrated that using recorded observations (data) will reveal changes that would not be noticed without someone having taken measurements. When we get to looking at case studies you will see that asking questions and challenging assumptions can help better understand a **system** and reveal potential undesired consequences.

With these core principles and the tools to use them, you have a foundation for examining how the environment reacts to changes brought about naturally and by human disruption of natural processes. You can use this foundation to identify a curriculum of study, examine proposed environmental management strategies, and interact with professionals involved in developing and implementing environmental management plans. For readers engaged in policy development or implementation or concerned about big environmental or **natural resource** issues, we provide you with a framework to use in evaluating options or strategies.

1.1 USING THE TOOLS

Predicting the consequences of actions and reactions beyond simple events that we can readily and repeatedly observe demands the recording of observations and often the analysis of these data. At times developing the questions to ask the data and assessing the answers can be more difficult than the observing and recording of data. This is because the question one asks leads to the methods to generate the data to answer the question. Bad questions lead to useless data. Good questions lead to useful data that help advance the inquiry.

As an illustration, let us consider the following example. Being told there is 25% chance of rain may not provide the information we want to know. Let us assume we know that the forecast provides the average probability of rain over the entire metropolitan area, but we know from experience that storms do not always spread evenly over our area. Thus, a 25% probability of rain for this forecast may not mean a 25% probability of rain for the specific area where we live. Perhaps we are more willing to wait out a short storm than carry an umbrella. In that case our concern is the duration of potential precipitation. Is it predicted to be a brief shower or a longer event? Or maybe we don't mind getting damp, but we don't want to be drenched. Then our interest is in the intensity of the predicted storms. In each of these cases we want know more than was provided by a general weather summary. The point here is asking better questions helps generate better information so you know whether or not to take an umbrella – in other words a better decision.

1.2 ASKING BETTER QUESTIONS

If we are to understand better the consequences of what appears to us as natural phenomena, especially where these do not seem to conform to our immediate experience, we need to know how to identify what questions to ask. Better understanding these events also helps us to avoid potential negative consequences (e.g. when to take action, what actions are enough, and do we need to modify our behavior). The point of asking better questions is to generate that amount of information to differentiate among choices. In most cases, the choice is between yes and no. In others, it may be short list of options. We generally only need just enough information to confirm or overturn our biases (predispositions).

Asking better questions means knowing what are the key factors that influence outcomes. In the following discussions of how and where formulating better questions is needed, we guide you to where to look and how to get better answers. This in turn will help you understand how factors influence outcomes and their relative importance, all of which leads to better decisions.

1.3 ORGANIZATION OF THE CHAPTERS

We begin in Chapter 2 with water and gravity. Water flows downhill, a cup full of water overturned will cause the water to spill. These experiences with our environment influence our notion of what is, and what will happen next. We do not need to reflect on whether or not a pail of water, when overturned, will cause the water to move, in what direction, relative to an incline, and how far. We know because we have witnessed the experiment, probably more than one time.

In Chapter 3, we expand the concept of everything goes somewhere and introduce you to the conservation of mass. Mass is neither created nor destroyed. What our senses do not tell us is that there is actually another way in which material may seem to disappear. For the vast majority of us, the phenomenon by which a chemical seems to disappear by reacting with another chemical element to form a new compound is relevant to our daily lives and our understanding of how our environment changes.

We explain the concept of conservation of mass with regard to the chemical elements – carbon, nitrogen, and phosphorus – their changing molecular forms and **physical states (gas, liquid, solid)**. We focus on carbon, nitrogen, and phosphorus for several very basic reasons: they are the key elements of life, and govern how much life there is. These elements determine how much food is produced and the extent of biologic activity. In our living world carbon, nitrogen and phosphorus are the limiting factors to growth. We explain how the three cycle through different pools in the environment (atmospheric, oceanic, terrestrial, biological, and soil), but within the constraint that the mass of each element is conserved.

Please note that figures presented throughout this book that may appear as point estimates for the amount of material in resource pools should be viewed as best estimates within large confidence intervals. In most cases, though multiple, well conducted studies have generated different estimated values, the relative sizes of the pools have remained fairly consistent. What is important is the concept, not the absolute number, since this can vary for a variety of valid reasons.

Chapter 4 introduces you to the principle of balance among cycling of the key ingredients of life. It is particularly in this realm that our intuition needs to be supplemented with additional information from several disciplines including soil science, hydrology, plant physiology, and ecology. Nitrogen, phosphorus, and carbon transform from one chemical form and physical state to another very different form and state, processes that are ongoing in the world around us although we observe little if any of its manifestations. The changes over time of their **concentrations** and relative proportions play a major role in how our environment functions. These relationships determine to a large extent how well we are managing our environment. We explain the role of soils, perhaps our least appreciated resource, in maintaining this balance. Using the framework of the agricultural environment, we illustrate how soils regulate water and **nutrient** cycles and influence the chemical interactions among carbon, nitrogen, and phosphorus. Soils are the locus of many of the critical processes that determine the availability of **bioavailable** nutrients, and water and nutrient storage.

Our discussion of soils brings us to Chapter 5 which addresses the question of how we have been managing our environment and natural resources. The focus for this book is the **agro-environmental** ecosystem that includes the land and soils of our agricultural land. Why agricultural land? Because much of our land is devoted to and affected by agricultural use. The productivity of our air, water, and soil resources depends on their quality and thus how these lands and these resources are managed. Because most of us live in urban settings we do not see the agro-environmental systems on a regular basis. For these reasons we examine how the agro-environmental system functions and explain how their management affects us all.

In Chapter 6, we introduce the reader to the natural and human-induced shocks to the **biogeochemical cycles**. We present a brief explanation of how the shocks take place and begin to trace the effect of these changes, introducing some of the consequences. These shocks are real and have occurred in our past, are occurring in our present, and will occur again in our future. At this point we assist the reader in using the tools that he or she has gained in the preceding chapters in predicting likely outcomes. Better understanding environmental processes is useful in itself, but we need to relate this science to the changing world around us to understand how change at various scales affects the environment. What happens when land use changes; additional carbon, nitrogen, and phosphorus are added into or subtracted from a system; or water flows are altered? These are outcomes that we read about in our daily news. This leads to the discussion of sticky, legacy problems of environmental management, the consequences of human-induced changes to environmental systems. Water and where it goes are also timely issues in the news. What newspapers and the news do not explain is how our environment affects whether or not water is simply an asset to humans or whether or not too much or too little leads to disaster. Gravity will bring the water in the form of rain or snow to earth, but where and how fast it moves upon the land will depend in good part on how we managed soils. We will explain how soils can serve as sponges sopping up excess water and allowing it to seep out slowly, moderating the impacts of excessive precipitation or flooding and minimizing the risk of too little water and hence drought. And water in the form of frozen water or ice can assist in the storage of vast quantities of excess carbon, nitrogen, and phosphorus, thus maintaining a delicate balance of the **reactive** forms of these elements on earth. Actions that can mitigate undesirable effects are also presented.

Chapter 7 introduces the concept of **'the Commons,'** a shared resource belonging to a group of people but owned by no one entity, and the issues that are associated with managing such a resource. Many environmental issues including water and air quality degradation can be traced to these resources being shared. This is followed by a set of case studies that illustrate instances where one or more of the core principles were ignored. The information and tools we have given the reader in the previous chapters should serve as a framework for assessing what could have been conducted to have avoided the problems. And by the way, for dollar values presented we use 2020 as the baseline year, unless otherwise indicated.

Finally, in Chapter 8 we end with a discussion of why we, all of us, should care. It would be remiss of us not to provide you with tools that can be used to improve environmental resource management without suggesting why you might want to use them. Public policy, that is the rules regarding how public resources are managed, varies with the governing system. Nevertheless, there are individual actions, consumption choices, and arguments for these actions that can be made regardless of how our public policy decision structure is constructed. We will discuss a few of these in this final chapter.

Humans will always alter their environment to make it work better, at least in the short term, and immediate to where we live. Understanding how nature will react can, with a little forethought, provide the opportunity to both minimize the adverse consequences of this reaction as well as to anticipate the outcomes so as to prepare and ameliorate their impacts. This can occur through both **mitigation** of the more consequential elements of our actions as well as adaptation to the circumstances. We will show that just a little bit of forethought can save us much pain. Or as our wise fathers used to say to us when we were teenagers, think before you act.

Chapter 2
The water cycle (hydrology)

All streams run into the sea, yet the sea is never full.
To the place the streams come from, there they return again.
— Ecclesiastes 1:7, King James version of the Bible (1611)

Let us start our examination of the environment with something with which you are familiar: the **hydrologic** (or **water**) **cycle**. The hydrologic cycle is an example we will refer to often that demonstrates the connectivity among the land, air, oceans, and biological systems (Figure 2.1). Rain falls from the sky and lands on the earth. The water either evaporates, flows over land, falls into lakes, rivers or seas, or enters the soil. Water from the rivers, lakes, seas, and land evaporates and returns to the atmosphere whereupon the cycle repeats. Surface waters also flow downstream from one place to another, as rivers flow into the sea. Water entering the soil either stays underground, flows laterally outwards (into surface waters), or is taken up by plants via roots and passes back into the atmosphere through the leaves, via **transpiration**. While there are a few other less traveled avenues, these are the major ways that water moves through ecosystems. The key point to remember is water does not disappear. It goes somewhere.

Let us start with the basic science and the factors that determine the flow of water that we observe: gravity, slope (of surfaces), roughness (of surfaces), vegetation (its physical structure), **permeability** (of soils and surfaces), intensity (of precipitation), and velocity (of flow).

© IWA Publishing 2021. A Guide to Understanding the Fundamental Principles of Environmental Management. It Ain't Magic: Everything Goes Somewhere
Authors: Andy Manale and Skip Hyberg
doi: 10.2166/9781789060997_0009

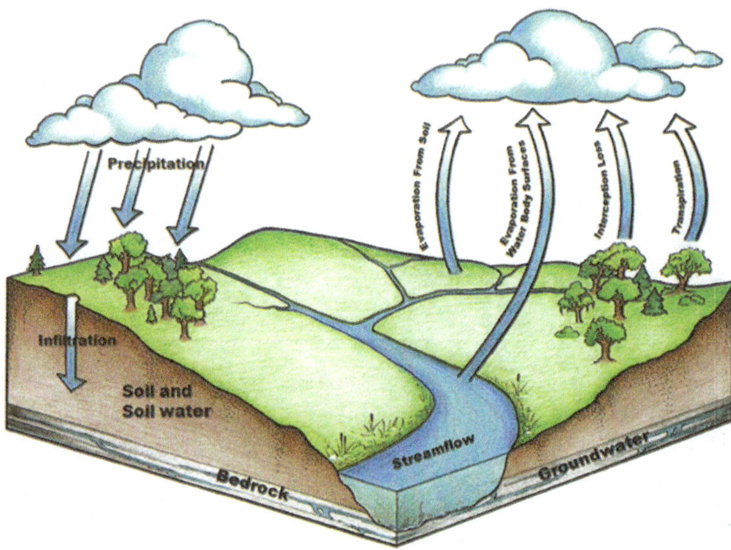

Figure 2.1 The hydrologic cycle – *source* Edwards *et al.* (2015). Drawing by Robin L. Quinlivan.

2.1 FUNDAMENTAL CONCEPTS THAT DETERMINE SURFACE WATER FLOW

Water, like Newton's apple, obeys the law of gravity. The law of gravity is unbreakable, and it does not take time off. It applies to water falling from the sky as precipitation, running downhill as surface flow, and infiltrating into and through the soil. This means that water will always move downwards, seeking the lowest point on the landscape, barring something blocking its path. [Things are a bit more complicated at the molecular level, which will be discussed briefly below.] Water can take many paths, but they all follow one principle – unless there is a waterproof barrier in the way, water will move from a higher place to a lower place.

How fast water moves (or flows) depends upon the slope of the ground. Surface flow is driven by gravity and modified by the slope of the surface. The greater the slope, the faster water flows down. Other factors, which will be discussed shortly, further affect the speed of flow. However, unless there is a barrier or other external force acting upon the water, it will move more quickly with increasing slope.

But what about the roughness of the surface upon which the water flows? We have all observed how water appears to flow more slowly down a rocky surface than a smooth one. A perfectly smooth surface presents no impediment to slow water flow, while a rough surface presents obstacles (e.g. ridges, stones, sticks, etc.) that inhibit and thus slows the flow. If plants are present, water moves more slowly across the surface. To think about this in concrete (so to speak) terms, imagine water flowing across glass, as compared to water flowing across a sidewalk. Which one disrupts the water flow more?

Experiment 2.1: During a rainfall, look at a driveway (paved) and a lawn next to it as the water flows onto the neighboring sidewalk. Compare the speed of the movement of the water down a concrete driveway against that across a lawn. If you measure the amount of water that moves across these two different surfaces, which part has more flow, the part next to the driveway or the part next to the lawn? You can also try to find somewhere that has areas with exposed soil and a good ground cover of plants next to each other. During a rainfall, which one has more water coming off of it? Observe how vegetation and vegetative debris – grass, trees, shrubs, leaves branches, and thatch – impede surface water flow and thus slow the water flowing downhill.

The plants impeding water flow demonstrate the effect of roughness, but when comparing areas covered with vegetation and those that are paved there is something more that is going on. Soil and surface permeability also impact surface flow. Permeability is the quality that tells us how much liquid (or gas) can penetrate into something. A porch screen, for example, is quite permeable. A drinking glass is not (hopefully!). We will discuss soil permeability in more detail a bit later, but for now it is important to know that water does not just flow over soils, but also infiltrates into them. And that water infiltrates into different soils at different rates and these soils can hold different amounts of water. For a highly permeable surface, water will flow into it instead of flowing across the surface. Conversely, the more impermeable a surface is, the more surface flow it will have. For our example looking at lawns and sidewalks, the reduced flow from the lawn is due both to the vegetation obstructing the flow and the **infiltration** of water into the soil.

Finally – but by no means the least important – how much water is going onto your surface affects how much water flows across your surface. The effect of this volume is moderated by all of the above factors – the cumulative impact of which is the ultimate determinant of how much surface flow, that is, the volume over time that you can have.

In summary, if you are looking at a surface and you want to understand how water flows across it, first recall what you already know – water flows downhill. Then understand that the rate of that flow will be greater with (a) greater slope, that is, greater angle; (b) less roughness; (c) less vegetation; (d) less permeability; and (e) increased flow intensity (like more rain, for instance). Now we are ready to examine another step in the cycle – **soil infiltration**.

2.2 WHY AND HOW WATER GOES INTO THE GROUND: SOIL INFILTRATION

First and foremost, water moves down due to gravity. Then it hits the ground and other surfaces. Those surfaces can and do have different permeabilities. If water hits an impermeable surface, like a sidewalk or a roof or a parking lot, it runs off that surface. This water flow is appropriately called '**runoff**,' or sometimes surface flow. The speed and amount of runoff are determined by the factors we mentioned above – the amount and intensity of precipitation, the slope, and the roughness. It is reasonably predictable based on those factors. When precipitation hits soil, however, there is a bit more to the story.

Soil may seem to us solid and impenetrable. We can walk on it, run on it, park cars on it, even build houses on it. However, as with so many things in this world, it is more complicated than that. Soil consists of minerals, **organic** matter, and a void, in other words, the empty spaces (pores) in between those solid particles. These spaces, or pores, allow water (and air) to infiltrate into the soil. For example, if we begin by thinking about a sandy beach, you have probably seen how this works. You can run on a beach – it is quite solid in that regard. But when a wave comes up and then down on the surface of the beach, the water does not run off, as it would on a sidewalk or parking lot. So where does it go? It flows into the spaces in between the individual particles of sand.

Soils are made up of different sized particles – the particles can be sand (larger particle sizes), silt (medium-sized particles), or clay (small to very small particle sizes). Different soils have different proportions of these particle sizes which in turn have different pore sizes between them. Moreover, larger pore sizes have more rapid movement of water through them. Experiment 2.2 in the following box illustrates this.

Experiment 2.2: Let us conduct an experiment that you can do quickly with household objects, with little effort. Get two cans and cut out both ends. Take two thin-meshed screens and put them on the bottom of each can. Then place the cans where the bottoms can drain. Fill one can with loose sand and the other with marbles or small stones. Now pour a can of water into each can. What happens? The water will flow through the can with the stones rapidly, while the can with the sand will release the water, but more slowly. An interesting side point is this: the can with the rocks has less pore space than the can with the sand.

If you do not have time or equipment to do the actual experiment, use the following conceptual exercise to fix the concept into your memory. Imagine two swimming pools. One is filled with bowling balls. The other is filled with sand. Imagine water falling on top of each of those swimming pools. Through which one is the water going to infiltrate more quickly?

2.3　WHAT WATER DOES ONCE IT ENTERS THE GROUND

After water has infiltrated into the soil, there are a few different things it can do. It can stay close to the surface or it can move downward. [Water remaining in the root zone of soils can be brought back up to the surface through **evapotranspiration**, the water goes into the atmosphere from **evaporation** and transpiration, which we introduce later.] Water moving downward goes towards the water table, the upper surface of saturated soil (Figure 2.2). There it can become **base flow** and move laterally through the subsurface or it can remain in long-term storage in an **aquifer**. The water table moving laterally as base flow can then come back out to the surface in rivers and lakes, where the water table intersects the stream channel or lake bottom. Subsurface water can also move laterally as it moves downward toward the water table. This occurs when the water reaches a subsurface soil layer that has become saturated and water cannot infiltrate downwards. The water will move laterally until it reaches a waterway or an unsaturated soil layer where it can continue its downward journey (Satterlund, 1972). Water in aquifers can be brought to the surface via wells.

In the ground, water will continue to descend, driven by gravity, until it meets an impermeable surface. Just as the water spilled from a glass stops its downward movement and begins to move laterally when it hits the floor, water in the ground moves horizontally via **subsurface flow** when it hits a less-permeable layer (Satterlund, 1972; University of California-Davis, 2020). A good analogy for saturated soil is a sponge saturated with water. When you pour more water on top of it, where does the water go? It comes out from the bottom and flows along the

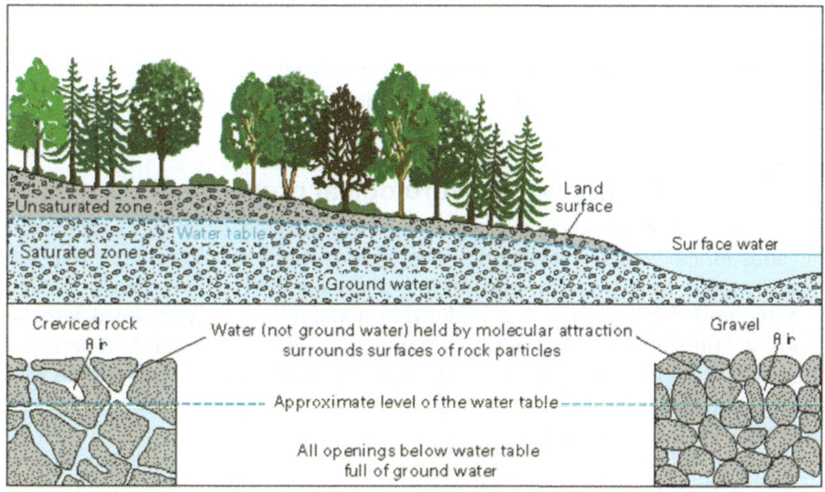

Figure 2.2 Soil infiltration and subsurface water – *source* Centers for Disease Control (2014).

floor. If you pour enough water, it comes out of the sides, causing the water level on the floor to rise. Similarly, water moves horizontally under the surface of the soil. Additional water moving down through soil moves because of the permeability of the soil. The direction of flow is determined by the path of least resistance, the side with more permeable soil.

If water goes underground, it is contained in the soil pores. It is not like there is some wide open space underground like a cave, holding all this water. [Well actually there are some underwater caves that hold water, but these are unusual and account for a very small portion of the water infiltrating into soils.] The soil pores below the water table are filled with water. Using what you already know, you probably have already figured out that the water table fluctuates as the amount of water infiltrating into the soil varies. In periods of high precipitation, it will rise and during long periods without precipitation or high **evapotranspiration** the table can fall. If the water table rises high enough (it can even rise to the surface of the soil itself), you can have a **wetland** – an area where the entire soil profile is saturated with water. Fluctuating water tables for a large number of depressional wetlands cause them to be wet during portions of some years and dry in other years. In the summer months, plant activity is high, causing evapotranspiration to use much of the available water. This reduces surface runoff, resulting in streams and rivers being fed mostly by this subsurface flow. Depending upon the physical structure of the underlying surface, the water we observe in streams and rivers may have fallen as precipitation days, weeks, or even years before.

Recognizing that underground water movement is not only downwards, but also sideways, is of critical importance for understanding surface water hydrology. Subsurface water moving sideways can come back out of the ground again. This **lateral movement**, or subsurface flow, is a critical source of water supply for rivers, streams, and lakes and is how **groundwater** provides base flow for surface waters. We will discuss this in more detail shortly.

Groundwater hydrological processes

We have provided in the text discussion a generalized simplified explanation of subsurface water movement. There are many complexities that we have ignored to provide a generalized understanding of how water moves underground. For example, the underlying geology can be inclined differently than the surface topography causing the water to move in a different direction than is indicated by the landscape. Although the water is in full compliance with the law of gravity, one might draw the conclusion the water is flowing over a hill. Impervious soil horizons can also play the same trick.

It should not surprise you that we cannot fully cover an entire discipline in a few pages. In situations where groundwater movement is an important consideration one should consult a **hydrologist**.

2.3.1 Lateral movement of water

There are three types of natural lateral flow: surface storm water flow, which is the water directly from a precipitation event; subsurface flow, which results from a storm moving more slowly than surface storm flow because it travels through the soil; and the base flow. What is base flow? Put very broadly, it is the lateral flow of groundwater. Base flow feeds waterways and other surface waters when the water table intersects with the stream channel or lake basin. It is the water that has made its way over time to the less **porous**, less-permeable rocky material beneath soils and, unable to sink further, pools and moves laterally, providing the water that you see in a stream when there is no runoff from storms. Identifying the point where subsurface flow from a storm ends and base flow begins is not cut and dried (Satterlund, 1972). For our purposes it is important to recognize there is a difference, but it is not necessary to identify precisely where it is.

Experiment 2.3: What do we observe when we dig a hole deep enough that water seems to appear from nowhere and partially fills the hole? When this happens, we have intercepted the water table. The water level in the hole is the height of this ground water. The water seeping out of the ground comes from the soil pores that are saturated with water. One place where this is easily observed is on the sandy beach at the seashore. Dig down into the sand until the water seeps into and fills the bottom of the hole. Here you have intercepted the water table. Now wait for an hour and watch the water level change as the tide ebbs or flows. The water level will rise as the tide comes in and fall as the tide goes out.

Now imagine that the hole is a pond – its bottom surface extends beneath the water table. For this reason, the subsurface water flows into it. This is also the case for streams, rivers, lakes, and other surface waters. Just to quickly note that this is a two-way street – a stream, for example, can gain water from the ground (called a 'gaining stream') and it can also lose water to the ground (called a 'losing stream'). The stream losing water occurs if the channel of the stream is above the water table (Baldwin and McGuinness, 1963; Chen *et al.*, 2013).

To visualize what is happening, take a good size length of clear plastic flexible tubing. Make a U of the tubing and then add water to one leg. Note that water rises in both legs of the U and observe that the water level in both legs of the tubing is at exactly the same height. Raise one leg of the U so you have a J-shaped figure. Again, the water level is the same in both legs. Return the figure to a U-shape and keep adding water into one end. When it is completely filled, the water will begin to flow out the other. Ground water behavior follows the same principle, although its behavior can be much more complicated. This is because different soils, rock formations, and other variables can create channels and barriers for water. See Figures 2.3a and 2.3b.

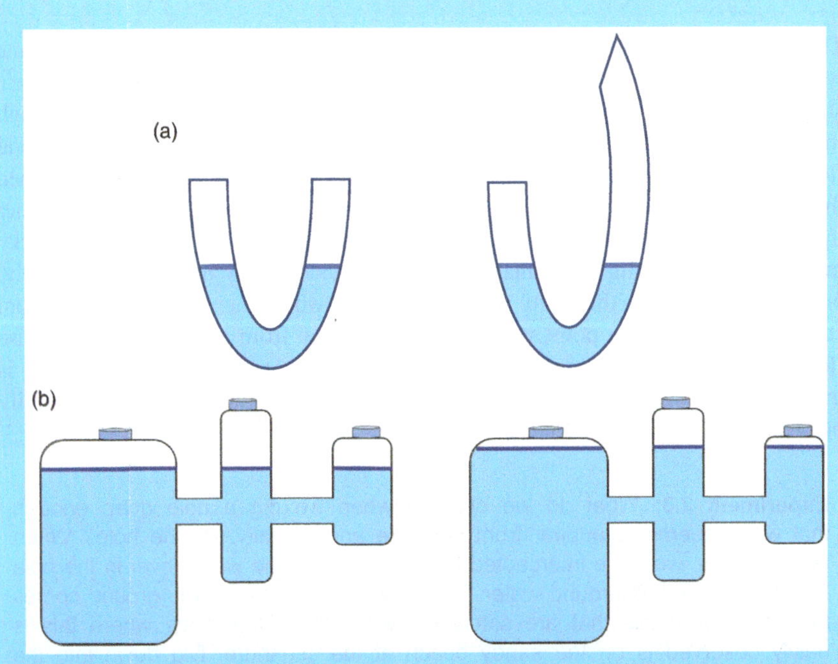

Figure 2.3 (a) Unobstructed water will flow to a single level. In the figure on the left-hand side is a flexible U-shaped tube with water in it. Note the water on either side of the U is at the same level. Now observe the figure on the right side. Note how one side is lifted making a J-shape, but the water remains at the same level on both sides. (b) Unobstructed water will flow to a single level. In the left illustration note that the water level is even in all three containers. Now add water to any of these containers. Note that the water level rises to a new level in each container and this level is the same in each container. If you try to fill the middle container you will be disappointed because the water will flow into the other two containers. If you squeeze one of the bottles, the effect on the water level is the same as adding water to a bottle, the level will rise in all three.

An interesting phenomenon is confined groundwater flow. This water, which results in what we know as a spring, still obeys the law of gravity but in a way that is not immediately apparent to us. Springs occur when rainwater that falls on relatively porous soil infiltrates until it reaches less-permeable soil or dense rock, then flows laterally through relatively porous rock that is contained within a less-porous rock formation. This groundwater flow, because the porous material that is sandwiched within the non-porous material acts like our flexible tubing, can make the water appear as if it defies gravity and rise from the depths. This happens when the source of the water occurs at a higher elevation to where the water returns to the land surface. We cannot see the rock formations beneath our feet. So it can appear as if the spring has no source. Because rock **strata** can be bent over long distances, without further sleuthing, you cannot know that the water may have originally fallen in the mountains far away from our spring.

How fast water flows underground depends upon the same factors as we have described for surface water flow. Clearly, underground there are more obstacles to flow than aboveground. Hence groundwater can flow extremely slowly, in some cases taking months if not years to reach the point where it rejoins the surface. In geologic formations of limestone known as karst, such as in the state of Kentucky and the Alps of Europe, the water can flow surprisingly fast underground. In these cases, the rains of winter can mean a flowing spring in the summer.

Subsurface, unconfined groundwater can be brought back up directly via transpiration or plants (as discussed in the next section) and via wells. Wells work, quite literally, the same way as our 'dig a hole into the water table' illustration, in our discussion of base flow. These days, they are generally associated with a pump that brings the subsurface water up. But if you see an old well, you might notice that it is a hole dug deep enough that it goes below the water table. A tube (often made of stone) is built to stabilize that hole. In this case, you might see a bucket that can be lowered into the hole to bring that water up.

2.4 SUBSURFACE WATER–PLANT INTERACTIONS

Water moves back up into the atmosphere through plant transpiration. Plants use roots to acquire water (and nutrients) from the soil. The roots absorb water from the soil pore spaces. The water moves through the vascular system of plants, which runs up through the plant, to the leaves. Here the water is released back into the atmosphere through small openings (called **stomata**).

Remember our experiment above with the flexible, transparent tubing. Water flowed out one end of the tubing when water was added in the other. The reverse can also occur. If you suck the water up and out the second tubing end, the level of the water in the first leg of tubing shifts downward. Likewise, water 'transpires' through the tubing of the plant, sucking up water from below.

And so, when we think of water moving through a plant, at one end are the roots; at the other are the leaves. The vessels connect them. The major driving force for this is evaporation at the stomata; the stomata are very small apertures, mostly in leaves, that allow for gas exchange with the environment. Underneath the stomata are small chambers, with cells in contact with the vessels. These openings allow carbon dioxide in (needed for **photosynthesis** and energy capture) and water out. When the water evaporates from these cells, they become more absorbent, drawing up water from the vessels (see **Capillary** Movement box).

Box on capillary movement – an exception
(sort of, but not really)

We have all observed capillary movement in action when we have dipped a cloth into water and watched the water move upward. Capillary movement is due to forces occurring at the molecular level. It involves two forces, water surface tension and the attractive force of water to solids such as glass, cloth, and the channels formed by soil pores. The surface tension is due to the greater attraction of water molecules to one another than to air molecules.

Capillary movement takes place in soils. When the attraction of water to the channel walls and surface water tension is greater than gravity, water will move upward within the soil. When the force of gravity equals these forces, the upward movement stops. The upward movement occurs in small openings where the combined forces can overcome gravity (see Figure 2.2).

The size of the pore determines how far water will move upward. Figure 2.4 illustrates the interplay between the attraction of the water to the sides of the tube and the gravity pressing down on the water. The water rises up to the point where water surface tension is strong enough to withstand the gravitational force. As the width of the tubes increases, the water level decreases.

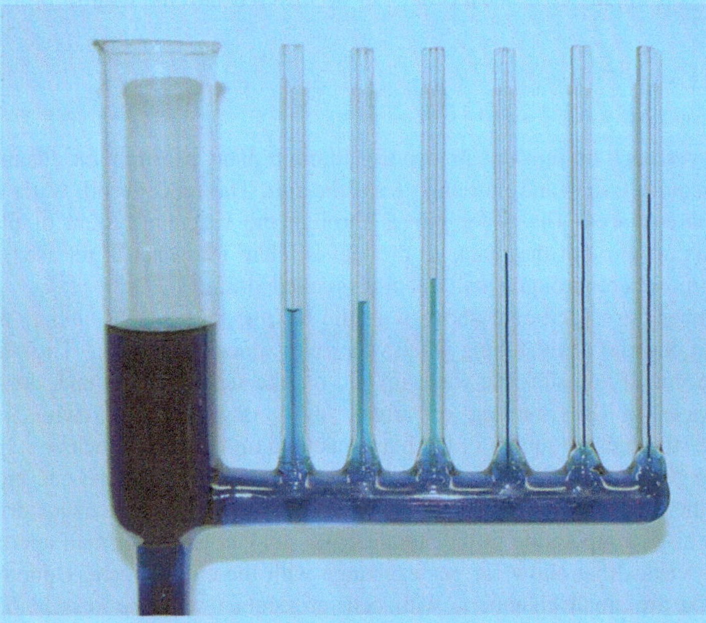

Figure 2.4 Narrower tube openings allow capillary action to pull water higher (Hayward, 2013 https://www.usgs.gov/media/images/narrower-tube-openings-allow-capillary-action-pull-water-higher).

> **Experiment 2.4:** How does water get up a tree? Exploring transpiration.
>
> Take a glass tube and a cotton ball, and, while holding them underwater, stuff the cotton (which is wet, because you are doing this under water) into the end of the tube. Take the tube out of the water and hold it vertically cotton side up. The water absorptive quality of the cotton and the attraction of the water molecules to each other hold up the water column in the pipette.
>
> Now take the cotton-topped-and-water-filled tube and put it in a jar about ¼-filled with water dyed with food coloring.
>
> The water gets drawn up (the evaporation at the cotton is analogous to the evaporation processes in a plant leaf) and you can see the colored water moves up the pipette.
>
> To clearly demonstrate that it is the evaporation from the cotton driving this you can do the following:
>
> (1)　Cover the cotton with parafilm or wax paper. You do not see the colored water move up (beyond the movement governed by diffusion).
>
> (2)　Use an empty tube. This illustrates that it is not **capillary action** that draws the water up – many people think capillary action drives water movement in plants. It does not.
>
> (3)　Take paper from a coffee filter, cut it into a leaf-like shape, with a 'stem/petiole' ('**petiole**' is the technical term for a leaf stem) and a 'blade' parts, stick the petiole part into the cotton tipped end of the tube and you can see how much faster the water movement is. This illustrates that leaves add evaporative surface and are not just for photosynthetic surfaces. They are also water pumps.
>
> (4)　Shine light on the paper leaf from #3. This also increases the water flow.
>
> (5)　You can also take the leaf from #3 and turn them parallel or perpendicular to the light. See how that affects water movement. It is a pretty marked change.

2.5 PEAK FLOW

We have already introduced base flow, which is the primary determinant of stream and lake levels between storms, but what happens during and after a storm? And what happens when the base flow is below the surface?

During a precipitation event, raindrops fall on open water areas, such as streams and lakes. Water levels rise. As we have discussed, the rain that falls on the land, if sufficient in volume, leads to surface runoff, and contributes to the rising level of the lake or stream. The peak flow is the highest water level attained by the water in the river or stream after a precipitation event. Depending upon the intensity, duration, and form of a storm the **peak flow** can include several different components (Figure 2.5 shows a graph of the water flow over time of **stream flow** (hydrograph)). If the precipitation is not sufficient to generate surface runoff or soil infiltration beyond the root zone,

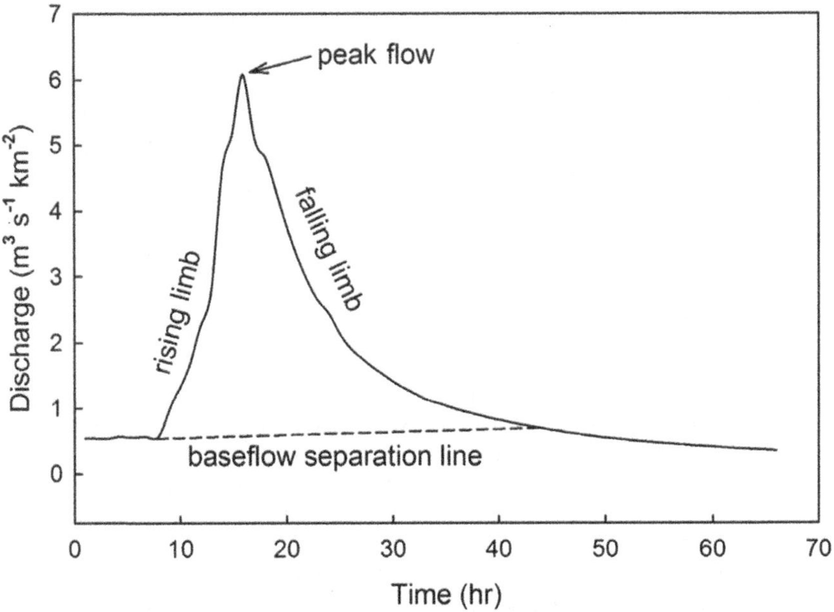

Figure 2.5 Stream flow before, during, and after a storm. Note the relationship between water flow in a stream from a storm and base flow. Stormwater enters the stream first as precipitation, then as surface flow, and finally as subsurface flow. The stormwater is in addition to the water from base flow. *Source* Edwards *et al.* (2015). Drawing by Robin L. Quinlivan.

the peak flow will consist of only the precipitation falling in the stream plus the base flow. The change from the storm will generally be unnoticeable. For a somewhat larger event, the peak flow will consist of the surface runoff added to the base flow. However, if the storm is of sufficient intensity and duration, the peak flow will include surface runoff, lateral storm runoff, and the base flow. The peak flow can occur well after the event if the precipitation fell as snow or in another frozen form.

The height and shape of the hydrograph (the figures given earlier) for a storm are influenced by a number of factors. The larger the **watershed**, the longer it takes for water to reach the main channel, delaying the peak flow. Larger watersheds also have more area so more water can accumulate from widespread precipitation. As we have discussed earlier, watersheds with steep slopes, impermeable surfaces, and fewer obstructions will move water more rapidly into the stream. These watersheds have higher and quicker peak flows than comparable streams with

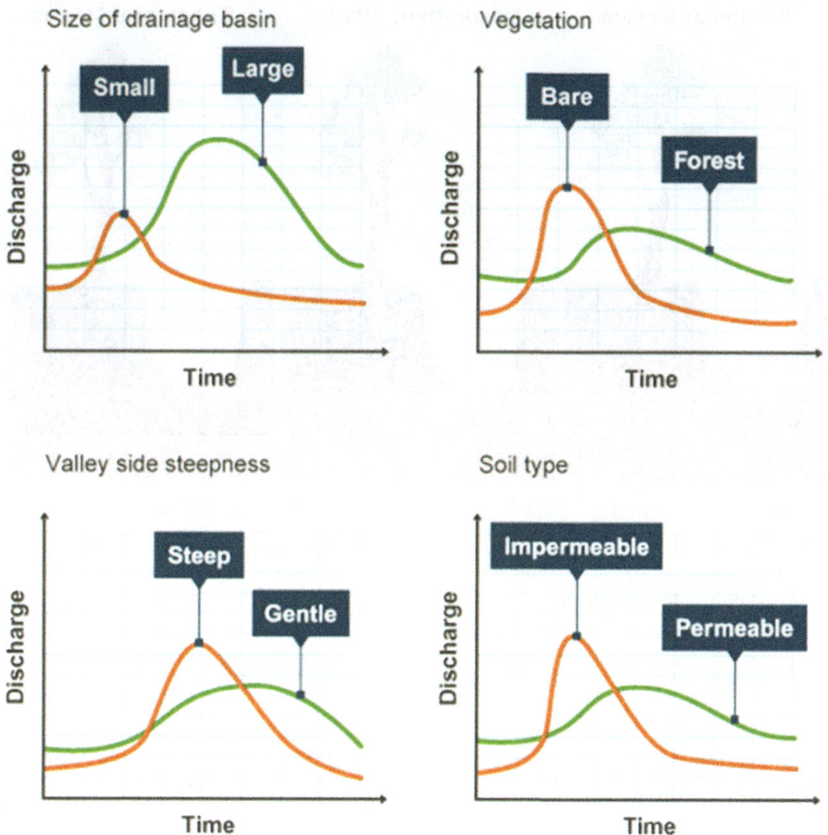

Figure 2.6 Hydrographs for watersheds with different characteristics (BBS Bitesize, 2020).

gradual slopes, more permeable surfaces, and fewer obstructions (Figure 2.6) (source BBC Bitesize, 2020).

A stream that flows all the time – it always has base flow – is a **perennial stream** (Figure 2.7). There are other kinds of streams, however, that do not flow all the time. **Ephemeral streams** only flow after there is a rainfall – their peak flow is their only flow. There are also **intermittent streams**, which flow seasonally when the water table is higher – their flow includes base flow. When the water table drops, they flow only intermittently; their storm flow is their only flow.

Water flow and the distribution of surface and subsurface water flow can be affected by people. This in turn can affect peak and base flow and ground water recharge. These anthropogenic influences will be discussed in Chapters 5 and 6.

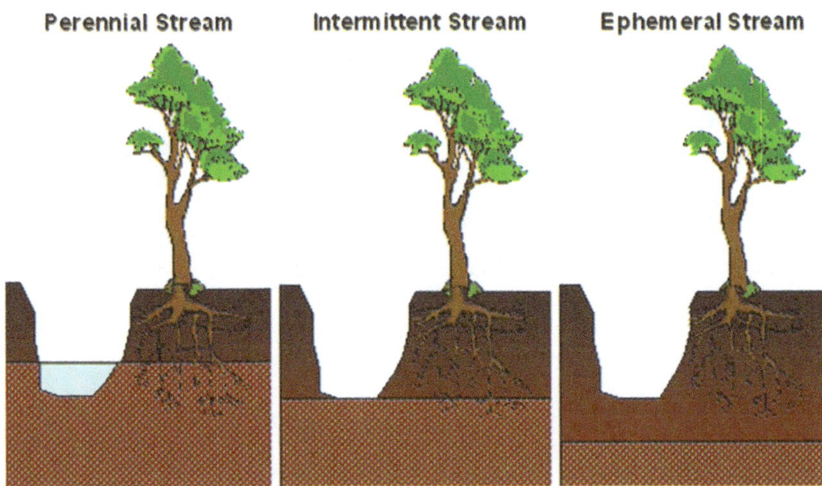

Figure 2.7 Relationship between base flow and perennial, intermittent, and ephemeral streams Zaimes *et al.* (2006). Note that base flow can fluctuate over time. The water in the intermittent stream will depend on whether the base flow is above or below the channel.

2.6 WHY WATER MIGHT NOT GO INTO THE GROUND: SOIL SATURATION

Impermeable surfaces are not the only reason water does not go into the soil. The soil pores can fill with water and become saturated. Then, when no more water can infiltrate, it flows over the soil surface. The following box describes an experiment that illustrates this.

> **Experiment 2.5:** Create a soil column in a tube, saturate it by pouring water on top until it stops going into the soil, watch water emerge through bottom of the core as enough water is added, and watch water flow off the top as the column becomes saturated.

2.7 WHY WATER MIGHT *NOT* GO INTO THE GROUND: CANOPY INTERCEPTION AND EVAPORATION

As anyone who has older siblings – or younger siblings, for that matter – knows, leaves hold water. A kid standing under a tree after a rainfall is just waiting for someone – like their older brother or sister – to come along and shake the tree and soak whoever is under it.

This is due to **canopy interception** – the ability for leaves, to retain the water that falls on them during a rainfall event. Not all of the water is caught this way, of course. Some of it continues through to the ground, as throughfall, and some of it runs along or the trunk and then to the ground, as stemflow. But a substantial amount of water can be retained as canopy interception. This water can then evaporate back into the atmosphere, or if there is an impact with the tree that shakes the leaves, the water may fall from the **canopy** (in which case it is throughfall).

Precipitation intensity also affects throughfall. A heavier rainfall leads to proportionately more throughfall than what occurs in a light rainfall. Comparing that amount of water falling under a tree to the amount falling outside of the canopy of that tree will give you the proportion that is throughfall. There is a large yellow poplar tree in my yard that I use as a bit of a rain gauge when I am reading in my office. When the leaves are out with moderate rainfall intensity, my sidewalk does not get wet until there is about a tenth of an inch (about 2.5 mm) of rain.

Water exposed to the atmosphere in streams, lakes, and wetlands evaporates, providing another avenue for subsurface water to move back up into the atmosphere. Soils that are not saturated also have evaporative loss of water as well, although at a much slower rate. It is through evaporation that the water that fell as precipitation returns to the atmosphere, completing the cycle.

> **Experiment 2.6:** Stand under a tree with a dense canopy during a light rainfall. Throughfall is about 0% and you do not get wet. Now stand under it during a heavy rainfall (without lightning for the sake of safety). Note that water drips through, meaning that there is throughfall.
>
> After a rainfall, shake a tree or branch and see how much water comes down on your head. Where would it have gone otherwise?

2.8 HOW DOES WATER GET BACK UP INTO THE SKY? EVAPOTRANSPIRATION

We have already covered transpiration – when water comes up through the roots of a plant, through its xylem, and out through the stomata. This is one avenue where water goes back into the atmosphere. Water can also evaporate from the ground, as has already been mentioned. Water can also evaporate from surface waters – this being the final major route by which the cycle leads back on itself, to where we began – water in the sky. Evaporation and transpiration, together, are called 'evapotranspiration.' To get a sense of the rate of evaporation from surface water, do the brief observational study in the following box.

> **Experiment 2.7:** Hold a glass or plastic tarp over boiling water or, on a hot day, over a pan of water. Note when you can first see condensation. Do the same on a day that is much cooler. Conduct this experiment multiple times with temperatures ranging from 4.5 to 32 degrees (40–90 degrees F). Is the relationship between temperature and the time you see condensation linear?

Evapotranspiration has a significant effect on how much of a rainfall event is actually available for runoff or infiltration. Notice after a heavy rain that the air appears heavy and sticky. It is filled – with water that has returned to its gaseous phase. In other words, it contributes to the **humidity** of the air.

How much of the water in a rainfall event returns to the air as humidity depends on a number of factors. Most important of these are aridity and **temperature**. Air in arid landscapes normally possesses low humidity. A rainfall event has the opportunity to contribute more moisture to the air before it becomes saturated. The point when air becomes saturated is temperature dependent, with air at higher temperatures being able to hold more moisture. The temperature where air with a given amount of moisture becomes saturated is called the dew point. If the atmosphere is saturated and the temperature is lowered moisture will condense. Temperature also affects the rate at which water turns into a gas. A hot surface (e. g. desert sand or city streets in summer) will transfer more energy to the water than cooler surfaces.

Evapotranspiration is also seasonal. Because much of the evapotranspiration is the transpiration component by plants and plants in temperate **climates** go dormant in the fall and winter, there will be less evapotranspiration in these seasons (Kirchner and Allen, 2020).

Experiment 2.8: You can observe the effect of lower temperature on saturated air during the summer months when the air is humid and the temperature drops at night. In the morning, you find the grass wet. What has happened is the lower temperature reduced the amount of moisture the air could hold, causing the water to condense. The dew point for a given amount of humidity is the temperature where air is cooled and can no longer hold all the moisture in the air and dew forms. Note that on mornings following very humid days dew is present even when the nighttime temperature is high. Conversely, on mornings following a day with lower humidity, a lower temperature does not necessarily cause dew to form. The United States National Oceanic and Atmospheric Administration (NOAA) provides a calculator that illustrates the relationship between the temperature, dew point, and relative humidity (https://www.wpc. ncep.noaa.gov/html/dewrh.shtml).

Experiment 2.9: You can observe the seasonal variation in evapotranspiration by comparing the persistence of puddles in vegetated areas during the summer versus the winter. During the summer when plants are active, the puddles dry up more quickly as the plants transpire and absorb soil moisture, allowing water to infiltrate into the soil. During the winter when the plants are dormant, the puddles persist much longer.

There is another important way that evapotranspiration affects runoff and soil infiltration and thus how much water is available for storage and for surface and ground flow. The direction of water flow (or flux upwards or downwards in the soil column) in the layer of soil closest to the surface is affected by the temperature of the soil and surrounding air. During the summer when temperatures are high, water can evaporate from wet soil faster than it can infiltrate. Hence, on net, water flux is upwards. In other seasons of the year under cooler conditions, the opposite occurs, water infiltrates soils (O'Geen, 2013).

Why is this important? As we will discuss in a later chapter, the direction of seasonal water flow matters in the situations where there is possible contamination, such as by **fertilizers**. For this reason, if we are concerned about groundwater contamination, we should focus our efforts to protect groundwater at times when water flow is downwards, that is, most likely in the spring or fall either before crops are transpiring or after they have matured and been harvested.

For most purposes we do not need to know exactly how much of the water evaporates at a given place and temperature. We only need to have a rough notion of what this value is in a discussion of surface water runoff, flood mitigation, or groundwater recharge. Surprisingly, the percentage of rainwater removed through evapotranspiration is generally quite high – on average some 50% (Rossi *et al.*, 2015). This value will of course vary according to season, precipitation quantity and intensity, vegetative cover, and temperature. Also, somewhat of a surprise to most people is the fact that the portion of rainwater available for surface runoff is generally only about 25% (California State Water Resources Control Board, 2011). In the hot summer months in the Corn Belt in the Midwestern United States, modelers generally assume an even lower value – as low as 5%. This number grows if there are sequential storms resulting in saturated soils.

2.9 WATER IN THE AIR – HUMIDITY

Water vapor in the atmosphere is mostly invisible, although you can see some of it as clouds and fog. You can also feel it – this is noticeable on muggy days. Some amount of water is always present in the air even if you do not notice it.

Water in the air is called humidity. Hot air can hold more moisture, cold air can hold less. The atmosphere is rarely at that maximum level. Relative humidity is the amount of moisture in the air relative to the maximum amount it could hold at that temperature expressed as a percentage [(amount of water in the air/maximum amount of water the air can hold at the current temperature) × 100].

At cooler temperatures, as can be found higher up in the atmosphere, water vapor can **aggregate** into droplets. These can fall as rain or as solids, such as ice or snow. Water falling as rain brings us back to the beginning of this chapter where we began our discussion of the hydrologic cycle.

When rain falls as ice or snow, there is a delay in its movement along the cycle. If you live in an area with snow, you see this in the winter. In places where the snow that accumulates over the winter and remains until warmer weather is called 'snowpack.' When it melts in the spring, it is appropriately called 'snowmelt.' If snow melts rapidly or in conjunction with rainfall, the resulting flow can lead to freshets and floods. Freshets are spring floods.

2.10 SUMMARY

The principles of hydrology can be understood through direct observations. You, yourself, can make most of these observations. You do not need to rely on complicated algorithms or even, for that matter, someone telling you what is true. Go out and look for yourself, keeping in mind water flows downhill and it goes somewhere. Keeping track of where water is going is often a first step in examining environment issues.

Understanding the basic principles of hydrology is an important start to understanding environmental issues. For any specific issue involving water volume or flow, you might be able to identify management proposals that are incomplete because they do not follow water fully through its cycle. Often examples of these inadequately designed proposals simply push a problem further downstream or underground.

Chapter 3

Conservation of mass or everything goes somewhere

Life plays Tetris with elements, and creates our reality.

— nebulaspage.org

We think there is color, we think there is sweet, we think there is bitter, but in reality there are atoms and a void.

— DEMOCRITUS, C. 460–C. 370 BC

In this chapter, we provide an introduction (we presume a refresher) to the core set of concepts critical to understanding the big, sticky environmental and natural resource problems of today. We describe the core set of chemical **elements** and molecules from which life and thus our natural world derives. We show how these elements conserve – neither increasing nor decreasing mass – though they may transform and change physical state (gas, liquid, and solid). We introduce these concepts because influencing or controlling where these elements go lies at the core of environmental management.

The key chemical ingredients of life, the chemical elements (**atoms**) that make up the building blocks that constitute all life, are also among the simplest and most abundant in the universe. But their abundance and hence relative availability does not predispose them to assemble into the components of life. To use an analogy, you may have all the ingredients for a cake, but just tossing them together does not make the cake. The recipe is important – they need to be assembled in the right proportions and the right order for success. This is where chemistry merges into biology.

Although the ingredients may be abundant, they are not unlimited. The atoms of each chemical element in this world, the physical world in which we live, are of a set number. Except through processes best described and explained by nuclear physics, they cannot be made nor eliminated. In every reaction involving these chemical elements, mass is conserved. The number of atoms of each chemical element that existed before a reaction exists after the transformation has occurred. For our purposes in our day-to-day world, the total number of atoms of a chemical element (say hydrogen or **carbon** or any other element in our natural world (see the periodic table of chemical elements) that exists is finite (Wise, 2013). You can, if you had the proper equipment and nearly infinite time, count each atom. Atoms connected (or bonded) to other atoms create molecules. Like the atoms, these molecules are finite – they can be counted. Were you to recombine atoms into different molecules or separate them into individual atoms, the original number of atoms of the chemical element will always be the same. This is the most fundamental and important principle of chemistry. This principle is a touchstone we can use to understand the world that we experience. Much of what occurs in environmental management is managing the chemistry to support the biology.

How is this concept important for environmental management? The answer is simple: it is key to understanding and resolving particular environmental or natural resource problems without inadvertently creating new problems that then have to be resolved – everything goes somewhere. In this and the subsequent chapters, we will provide many examples illustrating this fact.

Not all atoms or molecules can interact with other atoms or molecules to create new combinations. They have to be reactive under the conditions in which they exist. An element may be in a form that is not reactive (i.e., **inert**) because it exists alone or in combination with another atom in a way that requires large amounts of energy to separate. When the atoms of a chemical element (or a molecule containing more than one chemical element) are non-reactive, they do not combine with the atoms of a different chemical element. Hence, the reactivity of the atom of one chemical element can depend upon what chemical element is its partner.

On the other hand, an atom or molecule that is reactive may visibly change upon contact with another reactive molecule such that it changes its physical state (changes from a solid to a liquid or a liquid to a gas or vice versa). For example, add vinegar (acetic **acid**) to baking soda (sodium bicarbonate) and the two different sets of molecules will react, transforming into a third set of chemicals [see the chemical equation below]. The bubbles that form indicate that the carbon in the baking soda (a solid) has been transformed into carbon dioxide, a gas. Remember that the total number or mass of a chemical element stays the same after the reaction as before, as proven by adding up the individual atoms before and after the reaction.

$$NaHCO_3 \text{ (baking soda, solid)} + HC_2H_3O_2 \text{ (vinegar, liquid)}$$
$$\rightarrow NaC_2H_3O_2 \text{ (sodium acetate, solid)} + H_2O \text{ (water, liquid)} + CO_2 \text{ (gas)}$$

There were five hydrogen atoms (H), three carbon atoms (C), five oxygen atoms (O), and one sodium atom (Na) present before the reaction. After the reaction, the exact same number are present, though in different forms and states. Mass has been conserved. As illustrated by the earlier example, the atoms of a chemical element may change state when it combines with the atoms of other chemical elements. This changing of state along with chemical transformation underlies the dynamic nature of the basic elements of life. Understanding this is critical to the effective management of natural resources.

3.1 THE CHEMICAL ELEMENTS THAT COMPRISE THE BUILDING BLOCKS OF LIFE (AND MOST OF EVERYTHING IN OUR IMMEDIATE ENVIRONMENT)

The discussion earlier leads to the introduction to the chemical cycles involving the three chemical elements whose atoms comprise the building blocks of life. We explain their 'biogeochemical' cycling (cycling that involves biology, geology, and chemistry) from biological forms to various pools in the environment where they are otherwise stored. Each pool has its own timescale for how long the element typically resides in the pool. The different pools, the residence time in each pool, and the volume of flow from one to another make up the core functioning of environmental systems in both the living and non-living world. Remember all the while, even though the atoms of a chemical element have left one pool for another, the total mass stays the same.

Managing the forms of these elements and how they cycle lies at the heart of managing our natural resources and protecting the integrity of our environment. The atoms and their connections (bonds) with other atoms are often represented using ball-and-stick models. Each ball, depending upon what chemical element it represents, has a number of holes into which sticks are inserted to represent bonds. How difficult or easy it is to insert a stick into a hole of the ball depends upon the width of the hole and reflects the strength of the bond. In other words, if you exert little energy to insert the stick into the hole, then correspondingly the stick may just as easily fall out, breaking the connection. Too narrow and you must exert a lot of force to fit it into the hole, possibly destroying your creation in the process. One needs the connection of proper tightness to have a stable connection. That said, there are connections (bonds) of varying strengths and we will discuss these differences later.

This point is important because just as bond strength relates to the stability of the structure we build with our balls and sticks, how tightly atoms are bound to each other in molecules relates to their stability or reactivity in the environment. It also explains why certain chemical elements (carbon, nitrogen, and phosphorus) are ideally suited as the material that makes up the building blocks of life – **amino acids**, **sugars**, **lipids**, and **nucleotides**. [See Appendix C, Building Blocks of Life.]

Remember as children how when we wanted to build a house or a building, we started with pieces of wood that we cut and then nailed or glued them together to meet the requirements of our structure. The problem with this approach is that at the end of the day, we could not easily disassemble our construction or reuse its components. Once made, it was finished. On the other hand, if we used a LEGO^{TM} building block, we could use the pieces to construct whatever object our imagination conceived and at the end of the day disassemble them to store and reuse the pieces another day.

The chemical ingredients of life are the elements that comprise the blocks that are nature's LEGO$_S$^{TM}. And if you remember, there are only a small number of different types of LEGO^{TM} blocks, all serving different functions. The blocks have bumps (whereas the atoms have covalent bonds), when fit together, are **stable** enough to allow elaborate constructions. Yet they are not so tight-fitting that the blocks cannot be readily taken apart. And like nature, with LEGO$_S$^{TM}, whether assembled or disassembled, there is the same number of pieces and bumps before and after our construction projects.

The ingredients are carbon (C), nitrogen (N), and phosphorus (P), hydrogen (H) and oxygen (O). By assembling our ingredients of life – C, N, and P, along with oxygen from our water molecules (H_2O) – we create the building blocks. From these, we construct **proteins**; **DNA** (deoxyribonucleic acid) and RNA (ribonucleic acid), the carriers of genetic information; carbohydrates (sugars); and lipids (for cell membranes and fat, another means for storing energy). All forms of life as we know it are made of these units, from single-cell **organisms** to the complex plants, animal, and human world we know. As will be explained in greater detail, the carbon that starts as carbon dioxide in the air is transformed into a menagerie of molecular forms that comprise our biosphere. With a little ingenuity, multicellular organisms like ourselves can be created.

Obviously, there are other necessary chemical elements, such as sulfur, calcium, iron, magnesium, and potassium, but the discussion of these other elements is not critical to this story. Our focus is to show how using basic concepts for following and managing the chemical basis of our natural resources helps us understand our impact on the environment and many of today's sticky issues. These concepts can be applied to chemical questions beyond carbon, nitrogen, and phosphorus, such as toxic waste, nuclear materials, and plastics.

What follows is a discussion of each of the three chemical elements of life. We start with carbon and follow with introductions to nitrogen and phosphorus. [See Appendix B for a brief discussion of the chemical elements of life and their relationship to the early history of our planet and of life.]

3.1.1 Carbon

The chemical element carbon, our first key chemical ingredient, is represented as a ball that has four connection holes. Each hole or all of the holes can serve as a site to

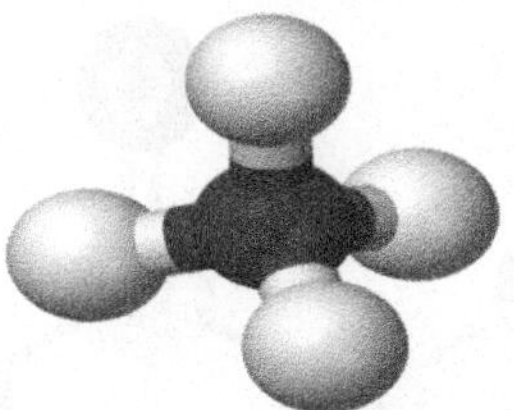

Figure 3.1 Ball-and-stick model of carbon with bonds to four hydrogen atoms – *Methane Image source*: Wikicommons.

fit a stick that connects one carbon atom with another or with a different type of ball, that is, a different chemical element. Carbon is ideal for serving as the framework for complex molecules that can be connected to each other to form long chains. It possesses the property of being able to connect with another carbon, or even three other carbon atoms in the case of diamonds, or with totally different atoms. The (covalent) bonds are not so tight that they cannot be broken when new connections need to be made, and their holes (representing the bonds) are not so weak such that the connections between atoms break too easily, making the structure unstable and easy to collapse. In Figure 3.1, we present a model representation of a methane atom – carbon (black ball) connected to four atoms of hydrogen (white balls).

Clearly if you are going to build a structure or many structures, it is handy to have a plentiful construction material. And there are a lot of carbon atoms in the universe. In fact, carbon is the fourth most common element in the universe (Suess, 1956). In all the different types of building blocks – amino acids, sugars, lipids, and nucleotides – carbon forms the core (Figure 3.2).

In Figure 3.3, we present the structure of one such common building block that you have heard about but might not know its chemical details. The molecule, with carbon at its core, is an amino acid. The O stands for oxygen and the N stands for a nitrogen atom. Amino acids connect to each other to form proteins.

So far we have not mentioned water. Water is important for constructing our building blocks. Why is that? Because constructing the building blocks requires a medium that brings the ingredients together. Water is a useful medium because it

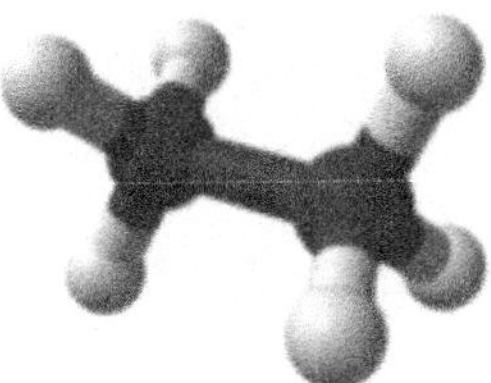

Figure 3.2 Butane: two carbons with H atoms. *Image source*: Wikicommons.

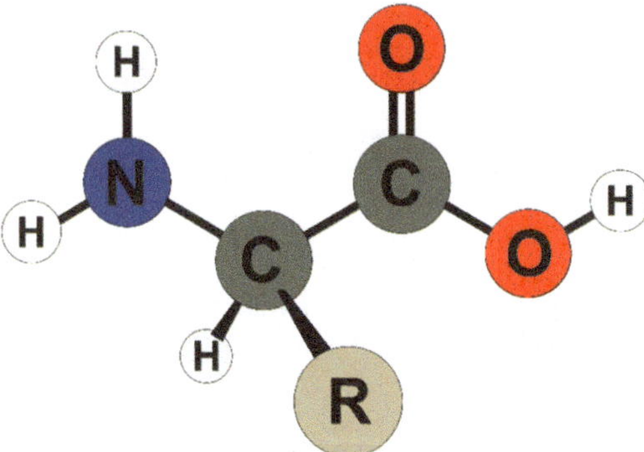

Figure 3.3 Amino acid. R stands for any of a number of atoms or molecules that can be added to the amino acid monomer. *Source*: Wikicommons.

is abundant and is liquid at the prevailing temperature of the earth at its surface, allowing atoms to be transported and interact. Moreover, it is a **dipolar molecule** that has a positive and a negative pole much like a magnet (see Figure 3.4). The significance of this is explained below.

As noted earlier, carbon makes a great starting point for constructing building blocks. But carbon–carbon chains are not **soluble** in water because of the **dipolar** (charged) nature of water molecules. Remember opposites attract and carbon chains have no charge. [Carbon chains act like oil in water – see Appendix A.] For carbon to be truly useful for constructing these blocks, it needs a partner to make it water-soluble. In other words, it must acquire a charge to interact with dipolar molecules like water. This partner, which must also be readily available on earth, is an atom that has a similar bond strength when connected to carbon; that is, the bond can also be readily broken. That partner is nitrogen.

3.1.2 Nitrogen

Nitrogen (N), the fifth most abundant element in the universe (Suess, 1956) meets the criteria needed to be carbon's partner. In the earth's biosphere, the region where

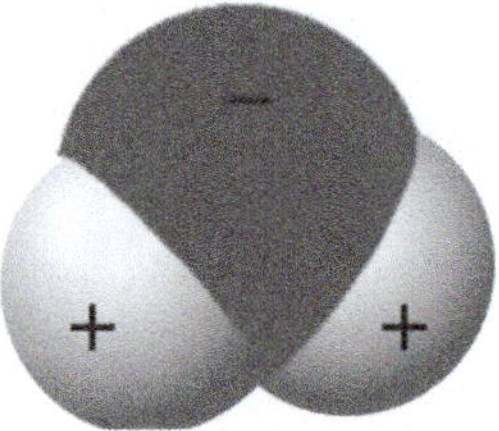

Figure 3.4 Water molecule as dipole. *Source*: Wikicommons.

Figure 3.5 Two N atoms as a gas. *Source*: Wikicommons.

life exists, nitrogen is abundantly present as an inert, two-atom nitrogen molecule [see Figure 3.5], but it is not abundant in a **reactive** form. It takes a good deal of energy to break the strong bond between the two nitrogen atoms, thereby making two highly reactive individual atoms. Once split, these atoms quickly bond with nearby atoms, such as oxygen or hydrogen, to form a relatively stable union, although much less stable than its two-atom form. In other words, the union can again be split if the right partner (with more energy) presents itself. With two bonds broken, nitrogen can connect with one and even two other atoms. This makes nitrogen ideal for constructing building blocks.

With carbon and nitrogen, we have two essential elements that when combined with oxygen (the third most common element), constitute our first building block, the amino acid [see Figure 3.3]. We can combine amino acids through nitrogen–carbon bonds, freeing up the oxygen to react with hydrogen and make water. [Because the interactivity of oxygen with hydrogen leads to the formation of water, nitrogen functions in the transfer of energy and facilitates chemical reactions. This applies especially to coupling and uncoupling actions – the making and breaking of bonds.] This queuing of amino acids results in a protein chain.

But this protein chain that is made up of amino acids still does not do work. In our example of the ball-and-stick model – now a chain of balls and sticks – there is no action. To do work, one must be able to store and transfer energy. For this, we need a mechanism for storing and releasing energy in a controlled manner–something analogous to a spring. A spring serves this purpose by being flexible enough that you can compress it, and the spring stores energy in its compressed state. When it is released, the spring returns to its original shape, releasing energy.

3.1.3 Phosphorus

Phosphorus, a chemical element never found alone because it is highly reactive, (i.e., it is always found in combination with another chemical element), plays a key role in the temporary storing, transferring, and releasing of energy to drive biological reactions. It naturally forms bonds with oxygen to form the compound **phosphate**. Phosphate contains four oxygen atoms and has a negative charge (represented as PO_4-). A single phosphate molecule can combine with its sisters into two and three phosphate forms, that is, di and triphosphate. The triphosphate form stores a biologically useful amount of energy. This is the energy that can be released, transferred, and used by biological systems to enable movement. More importantly relating again to our building blocks, phosphate can provide the energy to connect sticks to balls and thus assemble the building blocks that make up cells and hence life. The molecule to which the triphosphate form is attached, adenosine triphosphate (**ATP**), is known as the energy currency of life (Chemistry Libre Texts, 2020).

3.2 THE MYRIAD FORMS AND POOLS OF CARBON, NITROGEN, AND PHOSPHORUS

In the discussion earlier, we introduced you to the fundamental chemical ingredients that constitute building blocks: carbon, nitrogen, and phosphorus in their reactive or bioavailable forms. In this section, we discuss these key ingredients in detail to explain how they transform and cycle while conserving their mass. The discussion is important for environmental management because humans, in conducting economic activity, alter the cycling of some portion of the mass of these chemical elements or the mass that is cycled. In addition, this section provides a guide for you to use to unravel where the elements go when the cycle is disrupted.

3.2.1 Carbon

In the news, school discussions, family get-togethers, and elsewhere, the topic of conversation often moves to climate change and the need to do something about 'it' or the insistence that 'it' is a hoax or the outrage that there is even discussion of 'it.' This means a discussion about carbon, carbon dioxide, or some variant phrase containing the word carbon. Often, the concern or lack thereof is with the role of carbon with respect to climate change and what can be done or should be done concerning carbon emissions into the atmosphere to mitigate the perceived existential threat from 'it.'

Most people know that we exhale carbon dioxide (CO_2). And most people also know that carbon dioxide is taken up by plants. They may further believe that because we eat plants, carbon dioxide is good because plants need it.

Almost all carbon that makes up our biosphere, that is the biological world in which we live, has been recycled countless times through countless living organisms. One source of carbon is from long-past biological activity. More and more people know that **fossil fuels**, such as gasoline, coal, and in general petroleum products, contain this older carbon. In any case, it is important to understand that our biological world is made up of carbon in its myriad forms, which includes carbon dioxide, fossil fuels, animals, and plants. And carbon has

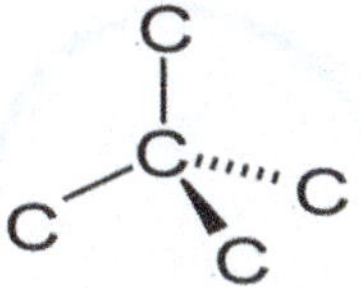

Figure 3.6 Carbon tetrahedron. *Graphite image source*: Wikicommons.

cycles, changes form, and is often transported great distances in its myriad forms, but always conserving its mass.

What is carbon? As we noted earlier, it is an element prevalent in the universe with four sites that can accept bonds, making it a versatile ingredient of building blocks. Like most of the chemical ingredients of life, it exists in our world in different phases – gas, liquid, and solid – depending not just on temperature, but also upon the other element(s) with which it is bonded in the molecule. In this regard, it is very similar to nitrogen and hydrogen. Carbon dioxide, a common chemical form of carbon, is a gas under normal terrestrial conditions.

Carbon (Figure 3.6) is the major structural element for life on earth in large part because of its capacity to form four strong bonds. When these bonds are between fellow carbon atoms in perfect symmetry to each other, carbon forms crystals with the greatest hardness in the universe – diamonds. Conversely, carbon that only joins with its lateral neighbors, in a plane, i.e., not in the tetrahedron of its sister diamond, forms one of the softest substances known – graphite (Figure 3.7), which is the 'lead' in pencils.

3.2.1.1 Carbon pools

Imagine a pail containing all Earth's carbon atoms, both as carbon and in combination with other atoms. The bucket represents a pool or reservoir of carbon (Figure 3.8) measured in **gigatons** (a billion metric tons or 1 Gt = 1,000,000,000,000 kg). Though there are some 2 billion gigatons (Deep Carbon Observatory at https://deepcarbon.net/) in the earth, let us use 50,000 gigatons for the amount in our bucket since most carbon lies deep in the earth and is inaccessible. Now imagine that the carbon in this bucket is distributed among five other buckets, not necessarily of the same size. The size of the bucket implies the size of the pool. [The numbers are derived from the Kansas State University Soil Carbon Center (2020). The values should be used for the purpose

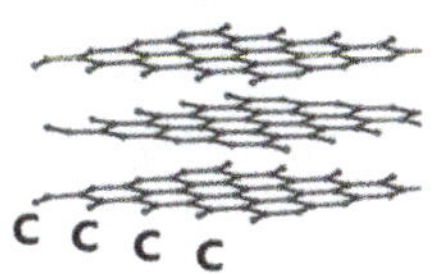

Figure 3.7 Graphite or hexagonal carbon. *Image source*: Wikicommons.

Figure 3.8 Total pool of carbon on earth, estimated to be some 2 billion gigatons, with slightly more than 50,000 gigatons near the surface. *Image source*: Wikicommons.

of comparison only because the amounts, particularly in the atmosphere, are changing over time.] Thus, we divide the mass of carbon near the earth's surface into five pools with differing amounts of carbon present in each bucket. Moreover, the carbon is also not necessarily in the same form. Some is pure carbon, some is present as carbon dioxide, and most is in a complex form with other atoms, such as proteins, soil **humus**, or fossil fuels. Nevertheless, the total mass of carbon across all buckets adds up to the same starting amount and hence we have conservation of mass (Figure 3.9).

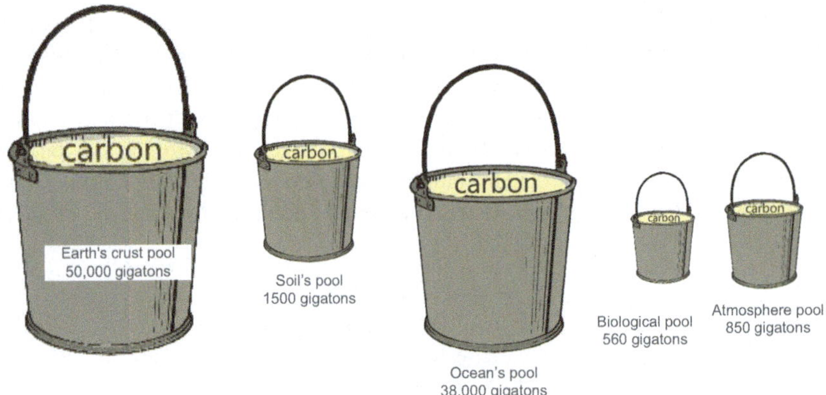

Figure 3.9 The important pools of carbon.

The last bucket denotes the carbon in the atmosphere, which is present as carbon dioxide (CO_2) and to a much lesser extent methane (CH_4) and carbon monoxide (CO). The total amount is some 850 gigatons as of 2020. Human activity increases this amount by about 6 gigatons per year. The bucket second to the right contains the biological carbon, which includes carbon in plants and animals but not the carbon in our soils. This organic carbon began as CO_2 and was then formed into simple sugars through photosynthetic activity. This bucket contains about 560 gigatons of carbon. The middle pail represents the 38,000 gigatons of carbon in the oceans. This carbon is mostly in the form of carbon dioxide in solution. The second bucket, the carbon in our soils, contains 1,500 gigatons of carbon. Finally, the first bucket, not drawn to scale, represents the carbon in the earth's crust and contains some 50,000 gigatons. Much of this is limestone or dolomite rock. This bucket includes the 5,000–10,000 gigatons of fossil fuel deep in the crust.

These buckets (or pails) are leaky (Figure 3.10), meaning some portion of the carbon in each bucket leaks out into one or more of the other buckets. The distribution of carbon among these pools is extremely important for the purposes of natural **resource management**. This is because it affects the balance of carbon relative to the other key elements of life – nitrogen and phosphorus – and it affects the variables of state (whether or not the substance is in a gaseous, liquid, or solid state), and the temperature, and the moisture (H_2O) content of land and air.

Figure 3.10 Carbon moves between pools. The boldness of the arrows indicates the relative magnitude of their importance.

The biosphere, biological pool (dead and alive), absorbs some 110 gigatons of carbon per year and releases back into the air some 60 gigatons. In this process, plants remove CO_2 from the air, convert it into sugars, which are converted into more complex carbon forms. As plants shed material and die, they fall to the ground, decompose, and are incorporated into the soil. Some of the carbon goes to the atmosphere as CO_2 and CH_4, and some 50 gigatons (the difference between the carbon absorbed and released by the biological pool) becomes part of the organic component of stable soil. Hence, a portion of the soils pool serves as a sink for carbon (humus), removing this portion of free carbon from its interplay with all the cycles. But the soils pool also contains a labile component, which can be lost back to the atmosphere bucket. In fact, the soils pool is currently absorbing on net some 1 to 2 gigatons of carbon per year less to the atmosphere pool than it did in the geologic past. The rate at which the soils pool absorbs less carbon depends upon how soils and in particular the land in general are managed. Conversion of forests, grass lands, or wetlands to an agricultural or urban use exposes the carbon in the soil pool to oxygen allowing it to oxidize and convert to carbon dioxide. The result is a net flux of carbon into the **atmosphere pool** from the soils pool (Rice *et al.*, 2009).

The oceans pool receives some 2.5 gigatons more carbon (generally as carbon dioxide) from the atmosphere than it returns. Some portion of this becomes incorporated into marine biota and, over time, incorporated into deep-sea deposits and **sediments**. These like soils, serve as a sink for carbon. This earth crust pool, which contains the vast majority of the carbon at the surface of the earth, diminishes as carbon is mined or drilled, and is burned and converted to carbon dioxide. This carbon is deducted from the earth crust pool and added to the atmosphere pool (Kansas State University Soil Carbon Center, 2020).

These flows among **carbon pools** are collectively called the global **carbon cycle**. What is not revealed by these graphics is the time dimension since this leakage from each bucket varies. The flow of carbon to and from the atmosphere pool and the biosphere pool (that is cycling) is rapid and occurs on a human **timescale** of days, weeks, and months. The carbon exchange between the atmosphere and the oceans is also rapid and occurs to a large extent annually, although some carbon gets caught in deep-sea currents and cycles over centuries. The flow of carbon between the biosphere and soils occurs annually as well, but the conversion of labile soil, as a consequence of a disturbance to these soils, can be rapid, less so for the more recalcitrant, deeper, more complex carbon component of soils. Finally, the fixing of carbon in the **crustal pool** is very slow on a timescale of thousands and millions of years. Human intervention, as mentioned earlier, changes this dimension. Thus, the carbon (mass) in each pool is the product of the interplay of biologic and geologic processes over a very long period of time, with the far greatest quantity being deposited in long-term storage. Carbon in storage is not available for use in the biologic world, making it relatively scarce.

3.2.1.2 Equilibrium between carbon pools

The carbon in these various pools is not static. There is constant movement between pools as carbon moves between the atmospheric, soil, and biological factions. Along with this movement, there are regulating forces that act to bring **equilibrium** in the exchange of carbon in these pools. This equilibrium has been observed over most of human existence. Up until the last several centuries, the amount in each pool has been balanced from the human perspective – the short geologic period of time in which human civilization has thrived – as carbon released into the atmosphere was largely offset by new soil production and mineral fixation.

We need to keep in mind that the different processes involved in each pool operate at different speeds. Any addition of carbon into the atmosphere is balanced by what is taken out through mineral formation, but these two processes operate on very different timescales. Large disruptions can last for long periods before the balance is restored. Carbon can enter the atmosphere rapidly with the **combustion** of fossil fuels. Rapid increases in atmospheric CO_2 levels result in rapid increases in the concentration of dissolved CO_2 in the ocean and other water bodies. CO_2 in oceans reacts with dissolved minerals as increased CO_2 in water increases its acidity. The CO_2 interacts with minerals, such as calcium, in the ocean, and this immobilization through **mineralization** is a slow process and measured in geologic terms – thousands and tens of thousands of years.

One way to think about this tendency toward equilibrium between carbon pools is the example we used for water. Envision several containers filled with water connected by tubes, the water reaching a constant level in each container (Figure 2.3). If water is added to one container, the level in all containers will reach a new equilibrium with the water level being the same but a higher level in each container. Similarly, if one of the containers is squeezed, shrinking its volume, the water level will rise in each container and again reach a new equilibrium.

For carbon this is similar, but there are two key differences. First, it is the proportion of carbon in each pool that reaches its equilibrium. If a system is at equilibrium and carbon is added to one pool from another, the system will adjust until the proportion of carbon in each pool returns to the equilibrium level. The second difference is that while water in our tub and tube example rapidly reaches equilibrium, the carbon in the different pools operate at very different time frames and reaches equilibrium more slowly. To put the difference in time frames in perspective, consider the fact that the tree you see outside and the oil and shale in the Permian reserve in the earth's crust are both parts of this cycle.

Let us reiterate that nature does not magically make the material appear and disappear. The carbon that appears to be lost from one state is just transformed into another state by changing its molecular alliances. Hence if you remove a unit of carbon from the atmosphere, this unit has to go somewhere. Most likely, it has cycled from its union with oxygen, where it is a gas in the air we breathe, to become part of a molecular complex in animals or plants. Given a longer period of time, it may become integrated into the soil. And over tens of thousands of years, it may become part of oil or coal deposits or even mineral rock. As explained earlier, the carbon in long-term storage can, at some time in the future, return to the biosphere as CO_2.

Ocean acidity

Oceans are a major sink for atmospheric carbon, as carbon dioxide. Absorbed by ocean waters, a portion of the carbon dioxide transforms into a compound called carbonic acid (H_2CO_3), a weak acid. Remember the discussion of conservation of mass; one water molecule combines with one carbon dioxide molecule: H_2O (l) + CO_2 (g) $\rightleftharpoons$ H_2CO_3 (aq). One or even both of the hydrogen atoms are so loosely bound to this new molecule that a portion of them dissociate into hydrogen ions (H^+) and bicarbonate ions (HCO_3-). The resulting freed hydrogen proton (the atom without its electron) changes the acidity of the surrounding water. [See Appendix A for a brief explanation of acidity.]

Ocean water is slightly alkaline (Rafferty, 2020; U.S. Environmental Protection Agency, 2020a, b). Up to the time of the Industrial Revolution, the pH measured roughly 8.19. Over the past two hundred years, the addition of carbon dioxide, largely though human activities, made it slightly less alkaline (pH 8.05), and thus 30% more acidic.

Changing the acidity within the environment triggers a cascading sequence of events. Any major change leads to major disruptions in the world around us. Acidity affects, among other things, the mass fluxes of chemicals, the chemical reactions that occur, and biological productivity. Each of these interact with others. Importantly for our discussion, it changes mass fluxes, i.e., the amount of CO_2, nitrogen, phosphorus, that enter one pool versus another and thus are biologically unavailable versus existing in a biologically available state.

Higher ocean acidity impairs the biological formation of calcium carbonate shells and structures causing certain ocean corals and the productive marine worlds they support to die. Higher acidity also leads to a corresponding decline in the rate that O_2 is absorbed by the oceans. This has a feedback effect because less O_2 reduces biological activity; hence, more CO_2 stays in the atmosphere, leading to warming as will be explained later. Higher temperatures also result in less CO_2 absorbed by the oceans.

Experiment 3.1: warm a bottle of soda – the soda contains CO_2. Watch the bubbles (CO_2) rise as the water holds less CO_2.

An example of a CO_2 reaction with minerals is the white crusting you might observe around your faucet that forms over time. If you are industrious in your cleaning, you may not observe this at your faucets. However, if your water has a high calcium content, the white buildup will be very evident. The interaction between CO_2 and calcium forms limestone. In the rock-forming process, this limestone becomes a longer-term sink for carbon.

Given prevailing conditions, the amount of carbon absorbed in water and contained in air will over larger periods of time come into equilibrium, and the change in flows between these pools nets to (essentially) zero. If the situation is changed by introducing more CO_2 (from stored carbon) into the atmosphere, the preexisting equilibrium is disrupted and the ocean absorbs additional CO_2 from the atmosphere to establish a new equilibrium to the system. In other words, the oceans serve as a 'sink' for the atmospheric CO_2. Over a long timeframe associated with mineralization of carbon, many disruptions of pools will occur and new equilibria achieved.

The properties of carbon in its different molecular forms play an important role in processes that can lead to disequilibria and also to more stable long-term equilibrium. We will talk in greater detail about events that can cause disruptions in later chapters. Here, we will introduce you to several processes involving carbon that play a role in how the carbon cycle reacts to these disruptions. Let us start with how atmospheric carbon dioxide and methane affect the **solar energy** entering our atmosphere and being retained by the earth.

3.2.1.3 *Effect of solar energy*

Solar energy, energy from the sun, either penetrates the atmosphere or is deflected out into space. If the solar energy enters the atmosphere, it can be absorbed by the atmosphere or penetrate to the earth's surface. If it reaches the surface, it can be absorbed or be reflected back into the atmosphere. The atmosphere can let the energy escape, absorb some portion of the energy, or reflect it back to the surface. A molecule that absorbs energy becomes excited – it moves around more. We perceive this as rising temperature. Changing the composition of gases in the atmosphere affects how much solar energy is retained at the earth's surface and thus temperatures and climatic processes.

Chemical compounds that increase the **retention** of solar energy are referred to as **greenhouse gases**. The major gases in our atmosphere, N_2 and O_2, which make up 99% of the mass of our atmosphere, do not have this property. They do not absorb infrared energy which is the wave frequency at which heat radiates (Mann, 2019). Carbon dioxide and methane are identified as greenhouse gases because they absorb infrared waves and hence prevent the escape of heat radiation. For this reason, all other things held constant, changing the concentration of CO_2 and CH_4 in the atmosphere changes earth's overall temperature. It is important to recognize that there are many interactions and feedback mechanisms that also influence the earth's temperature and climate. As we noted earlier, these processes operate at different rates, so a process that causes a disruption by increasing atmospheric carbon may not be offset for long time.

Plants and some **microorganisms** capture solar energy through photosynthetic reactions and use it to build the building blocks from the chemical ingredients. Over time, this captured solar energy is stored in soils or, in the case of the ocean pool, in detritus that sinks to the lower regions of the ocean. Soils, in turn, over even greater periods of time, are mineralized and converted to rock. In the **terrestrial pools** (soils and plants/animals buckets in Figure 3.9), the organic matter can even be reduced to complex **carbonaceous** forms that we call fossil fuels. In the oceans, the carbon in detritus is transformed into chemical species that serve as long-term sinks for carbon. Limestone is the product of very long-term accumulation of carbon-containing compounds and sedimentation in the oceans. This latter process occurs of course, exceedingly slowly over thousands and millions of years. Thus, over a very long period of time, the crust serves as a sink for the carbon in the terrestrial and ocean pools.

With an increase in carbon, as carbon dioxide and methane, in the atmosphere more solar energy is retained as heat which leads to higher temperatures. Higher temperatures, in turn, affect that portion of the atmosphere in which we live and work. Higher temperatures affect the cycling of carbon through the terrestrial pool with the mass of carbon in this pool declining and the amount lost returning to the atmosphere. Higher ocean temperatures lead to a lower rate of absorption of CO_2 in the oceans. Whether or not plants will increase production (and thus increase their mass of carbon) by absorbing more carbon as a consequence of the higher CO_2 level – thereby reestablishing a carbon balance within the system – is unclear (Wong, 2020).

3.2.1.4 How disruptions to the equilibria among pools can occur – atmospheric carbon emissions

Now that we have discussed carbon pools and sinks, let us examine some of the ways disruptions redistribute carbon, most often by returning carbon to the atmosphere, and consider how the disequilibrium they cause can impact environmental systems.

Combustion: When most of us think of combustion, fire comes to mind. Fire is certainly a prominent form of combustion, so let us start with fire. Light a match and feel the heat. The matchstick blackens and diminishes in size as the combustion causes the carbon to escape. As you are probably getting tired of being reminded, the carbon did not disappear, it transformed. The combustion, fire in this case, supplies energy that results in the breaking of carbon bonds in wood and replacing these with bonds with oxygen. The gas carbon dioxide is formed and then escapes into the air.

Fires that occur naturally or part of ecologic cycle: A corollary to combustion is the occurrence of forest and grassland fires periodically in the lifecycle of **ecosystems**. The great American conservationist Aldo Leopold (Tanner, 2012) described how fire plays an important role in preserving soils and hence restoring the basis of ecosystems, particularly in the more arid western regions of the United States. Over time, carbon and hence also the nutrients associated with the carbon build up in plants and dead matter. Wood in trees and shrubs can store large amounts of carbon. By releasing the carbon stored in vegetation, nutrients are freed to return and replenish the soils. Combustion shifts carbon from the terrestrial pool to the atmospheric pool momentarily and enriches the soils enabling vegetative growth to recapture the carbon and return it to the terrestrial pool. Overtime, a portion of this carbon in the terrestrial pool moves to longer-term storage, thereby sequestering and removing it from short-term cycling (National Geographic, 2020).

The difficulty of restoring ecosystems altered by humans

In post-industrial societies around the world, there has been a movement to restore lands to their original ecosystems or preserve them in what is believed to be their original state. Perhaps the most telling example is the emerald green peatlands of Ireland. As ecologists set about establishing standards to protect certain 'pristine' lands, they discovered that the almost treeless landscapes that most Irish consider original (or unaltered by man) and that coincidentally appear almost as putting greens – thereby occupying a space in the national character – are not virgin. In fact, the lands had originally been forested. Some time long ago, the trees were chopped down. The result was the deep green peatlands we know today. Thus, the question becomes the following: to what do you restore a human-altered landscape? The question is not one that science can necessarily and easily answer. Science can more readily tell us what ecosystem characteristics are possible given climatic circumstances. Indeed, social preferences play perhaps a more important role in decoding the dilemma.

Combustion and fire

We have all watched a fire consume the paper, kindling, and wood that once filled the fireplace, and leave a small pile of ash. This example illustrates how mass is conserved, though changing form and state. We know that the carbon in the combustion material interacts with the oxygen in the air to become the gas carbon dioxide. Not as easily observed is the combining of oxygen with nitrogen in the air, though only at high temperatures, to make nitrogen oxides, other gases. Both reactions are made possible because the energy released by the combustion broke bonds between the nitrogen and oxygen atoms in the N_2 and O_2 molecules. This freed the atoms to combine with each other and carbon. As the carbon in the combustible material is used up and the fire eventually quenched, the result is a shift of mass in each pool (biologic to atmospheric) through molecular recombination. Overall, mass is conserved.

Land cover change: Different land covers process and hold different amounts of carbon. Changing these covers from one to another changes the carbon stored and processed. When a cover such as a forest or wetland that contains relatively large amounts of carbon converts to rangeland or cropland that holds smaller amounts, carbon is released, generally into the atmosphere. Over short periods, the amount of carbon absorbed from one land cover can vary significantly. Over long periods, the changes in the amounts that are absorbed into the terrestrial pool can be quite substantial and cause disruptions (disequilibrium) between the terrestrial and atmospheric pools. Consider for example the change in the mass of carbon in the terrestrial versus the atmospheric pools during the ice ages or the tropical conditions that existed in the Jurassic Period (Tang, 2020).

Volcanoes: Volcanoes bring to the earth's surface materials containing carbon, generate CO_2 emissions, and ignite forest and grassland fires. The fires that occur from eruptions are normal occurrences in ecosystems around the world. Over millennia they have resulted in CO_2 emissions that roughly canceled the amount of CO_2 that was sequestered by plants and microorganisms on land and in the seas, leading to a net CO_2 flux that was generally zero or slightly to the benefit of sequestration. This slight net absorption of CO_2 by plants relative to emissions by fires over a long period of time explains, in part, the slow decline in atmospheric CO_2 during the periods between **ice ages** that has been detected by an examination of ice cores and other physical earth records. It also explains why scientists once postulated that the earth was headed toward a new ice age.

Oceanic versus terrestrial photosynthesis: How important is the land versus the ocean in fixing carbon? This is not an easy question to answer. One cannot easily go out on land and the sea and measure plant and soil growth. At best we can use models utilizing data on ecosystems, the extent of ecosystems, and limiting factors, such as nitrogen, phosphorus, and water, and temperature, to develop feasible scenarios and estimate the corresponding answer. One hypothesis is that photosynthesis on land and sea is roughly equivalent (Field *et al.*, 1998).

3.2.1.5 Natural constraints on productivity

If we remember that life follows a recipe that must be strictly observed, we can surmise that the most productive areas are likely to be areas with ample water, nutrients (whether coming off the land or carried by rivers), and temperatures that do not freeze cellular activity. Coastal **continental shelves**, **estuaries**, regions with high rainfall, and wetlands each meet these criteria, and the continental shelves, estuaries, **floodplains**, and wetlands are all extremely productive because they have all these ingredients including temperatures that are neither too hot nor too cold. For this reason, they convert large amounts of CO_2 into organic carbon. Wetlands and floodplains can sequester some quarter metric ton of carbon per **hectare** per year (223 pounds per **acre**) (Lal *et al.*, 1998a). Drawing on an analogy from our children's fairy tales, these highly productive systems have Goldilockean qualities. On the other hand, much of the earth's surface is not nearly as productive due to one of several limiting factors. Although they cover most of the earth's surface, much of the ocean area is not very productive. Even though CO_2 is readily available, and there are numerous locations where there is ample phosphorus, oceans tend to be low in nitrogen (Smith, 1984; Zehr & Ward, 2002). The scarce nitrogen limits the construction of life's building blocks and thus the propagation of plants and microorganisms. Terrestrial systems, on the other hand, tend to be constrained by water as well as limited amounts of biologically available nitrogen and phosphorus (Lal *et al.*, 1998b). The productivity of rivers, except at their headwaters, tends to be limited by a lack phosphorus.

3.2.1.6 Role of temperature in the carbon cycle

Temperature matters both to the form of carbon, but also in what pool it is found (i.e. in the ocean, biological, atmospheric, soil, or crust) and to its availability within the biosphere. Sufficiently cold temperatures that continuously freeze land, as in the northern hemisphere's boreal region, prevent the decomposition of plant matter. The resulting **permafrost** soils accumulate organic carbon from the growth cycle of each year. Soils in these regions can be very deep, containing large amounts of carbon drawn over time from the atmospheric pool (Sherburne, 2020). Thus, these soils store carbon that otherwise would occur as carbon dioxide and reduce the amount of CO_2 in the atmosphere where it would have acted as a greenhouse gas. Temperatures that rise to the point that permits these permafrost soils to defrost result in the stored vegetation decomposing and releasing the CO_2 back into the atmosphere. Although the warmer temperatures will increase photosynthesis and the carbon contained in this year's vegetation, the net effect of the higher temperature is to transfer carbon into the atmosphere – because the carbon released from the decomposition of the previously stored vegetation overwhelms the carbon captured in the vegetative production of the current year.

In temperate regions, plants have adapted to the current temperature conditions over tens of thousands of years. A significant shift up in temperature moves many if not most plants, out of this sweet zone, lowering productivity, reducing plant growth and yield. Moreover, the effect of temperature on climate (and thus on water availability due to changes in evapotranspiration and perhaps precipitation) means that this one key ingredient for biological action changes, altering the conditions for plant growth. Even with higher CO_2 levels which theoretically could promote plant growth, the absence or surplus of water, as we have seen, a key ingredient in biological activity, impedes growth. Over thousands of years, plants can adapt by either migrating or evolving. In general, higher temperatures lead to less organic carbon in the biosphere. Systems adjust to these new circumstances by losing soil carbon. This is precisely what has been happening in recent years in the boreal forests of Siberia, Canada, and Alaska with constant forest and peat fires, that is combustion and the generation of CO_2 from what was carbon in the biological and soils pools (Sazonova *et al.*, 2004; Schuur *et al.*, 2018).

In areas with a greater range in climate conditions, such as California, seasonal wetness leads to explosive plant growth and hence carbon capture which in turn is often explosively released in grass and forest fires. Soils and the vegetation have evolved to readjust to lower water availability in drier seasons. Over millennia plants have adapted to these extremes by storing in the soil the excess carbon accumulated in good years. The aforementioned plant burns, but the roots or seeds persist allowing for regrowth when water is again available. The carbon serves like a sponge holding in water. Ever rising temperatures, however, lead to ever declining levels of carbon in these soils. The sponge gets smaller holding ever less water to help the plants withstand weather extremes.

In the ocean, one can witness a different effect but with the same impact on CO_2 levels in the atmosphere. Colder water holds more dissolved carbon dioxide. One can also witness this phenomenon in our experiment with carbonated water. As the water in oceans gets warmer, its capacity to contain carbon dioxide decreases (National Space and Atmospheric Administration, 2019). Even though roughly a third of new CO_2 emitted into the atmosphere gets absorbed by the oceans, ocean currents can result in net emissions as warmer temperatures cause upwelling and mixing of ocean layers. In combination with the higher acidity of water from higher atmospheric CO_2 [see textbox on Ocean Acidity], the conditions for ocean microorganism growth changes. Phytoplankton, which are located at the bottom of the ocean's food chain, become less productive. Less productivity means less carbon fixation even as more carbon is being absorbed.

3.2.1.7 Summary

In telling the story of carbon, we have to reiterate that our biological world centered around carbon is a larger system composed of interlocking smaller systems. All parts of the system are important for the health of the whole. Removal or disruption in one part has consequences for all other parts of the system, although these effects may occur at varying temporal and spatial scales. Witness how higher organisms such as birds, buffalo, and salmon help distribute nutrients and enrich the system (Gende *et al.*, 2002). Carbon in the biological pool above ground, such as trees and grass, feeds the soil carbon pool below ground. A change to one affects the health and viability of the other. Carbon in the air cycles with carbon in oceans and terrestrial systems. Temperature and water are affected by what happens to the carbon in all its forms.

3.2.2 Nitrogen

The story of nitrogen (N) provides a good introduction to how single chemical elements can assume many forms with vastly different properties and ecological consequences. It illustrates the many chemical paths an element can take. It also shows how humans can alter natural cycles through innovation, in this case an artificial fertilizer for crops and the intensified use of **leguminous**, that is, or nitrogen-fixing crops. And it explains how disturbance in the balance in one chemical's cycle impacts those of others.

You breathe nitrogen in every minute of every day, yet the nitrogen you breathe is not absorbed into your body, but is expelled in the same amount that is inhaled. It is the nitrogen that you eat and must consume in significant quantities to live, which is taken in by your body. This nitrogen you get from **legumes**, and especially meat, fish, and dairy products. It is what, in one form, is severely limited in quantity in nature and, in other forms, present in vast, almost unlimited, quantities.

Nitrogen as N_2 makes up some 78 percent of the earth's atmosphere, but only a small portion of nitrogen is biologically available, because as N_2 it is **inert**. If you remember the example of the balls and sticks (Figure 3.5), N_2 is represented by two balls that are tightly bonded to each other. Natural processes involving large amounts of energy (e.g., lightning and combustion) and biological fixation break the strong N_2 bonds to make nitrogen available in a reactive form (henceforth referred to as rN), literally out of thin air. As a plant nutrient, rN supports plants in grasslands, forests, and oceans. These plants are food for **fauna** and, as food, supply rN to animals. Plant material when it is shed or dies, delivers **rN** to the soil where it can be recycled and nourish other plants. With rN available, life blooms because it is generally the limiting factor for growth. Excess rN,; however, causes fragile ecosystem systems to fail.

The locking up of rN in soils or in minerals on land or in the ocean in complex compounds prevents its release into air or water as reactive forms. If sufficiently buried in the soil or if the land is not disturbed, these latent forms of rN are unavailable to interact with other elements that would return them to a more reactive state. They exist as organic matter in soils or in isolated conditions deep in water bodies, such as oceans and lakes. Nitrogen is also locked up in chemical complexes with minerals deep in the earth. There is a time dimension to the question of reactivity or availability. Today, nitrogen may not be available to react. In the future, environmental conditions, such as a volcanic eruption, flood, drought, or earthquake, may cause it to be again 'reactive.' However, only a major natural disturbance results in their release and conversion. Hence, in any discussion of nitrogen be aware that there may be agreement with regard to the science of nitrogen but in disagreement with regard to the time scale in which the nitrogen availability occurs.

3.2.2.1 Nitrogen pools

Nitrogen, like carbon, is stored in pools. Unlike carbon, over 99% of all the mass of nitrogen is present in the atmosphere as the gas, N_2 (Figure 3.11, Table 3.1). [There are a wide range of estimates for nitrogen stocks, however, there is broad agreement concerning the relative values of the of the amounts in the pools.] Less than one percent of nitrogen is stored within the earth's crust, dissolved N_2 in ocean waters, or as non-organic compounds in the oceans. A very, very small percentage is present in soils and an even smaller percent is present in **biomass** in the biosphere. Only some 0.005% of the mass of nitrogen is reactive, that is, available in a form that can react with other atoms and can be incorporated into life. In other words, nitrogen is scarce in nature in forms that can be incorporated

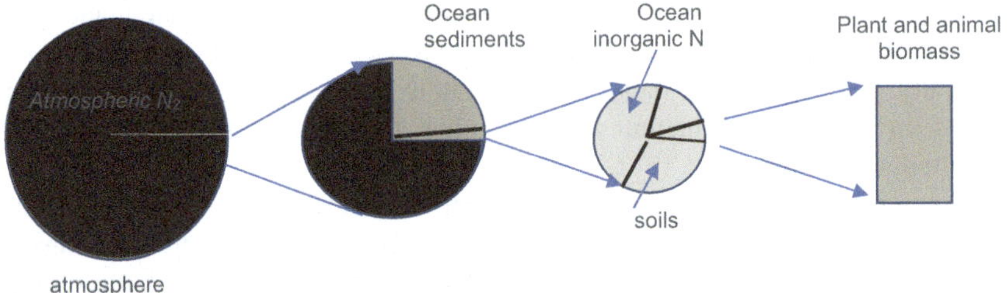

Figure 3.11 Nitrogen pools.

Table 3.1 Nitrogen Pools.

Non-reactive nitrogen	Atmospheric (N_2)	99.3245%
	Sediments (crust)	0.54%
	Ocean dissolved N_2	0.1242%
Reactive nitrogen	Ocean inorganic N	0.0015%
	Soil	0.0024%
	Terrestrial biomass	0.0009%
	Atmosphere N_2O	0.0003%
	Marine biomass	0.00006%

Source: U.S. Environmental Protection Agency (2009).

into living organisms. This has implications for natural resource and environmental management, because humans have, since the early twentieth century discovered how to overcome this scarcity and create new reactive nitrogen from inert nitrogen.

Most of the rN is not in a form that is bioavailable (Table 3.1), and the great majority of rN is locked up in soils (in complexes with carbon and phosphorus). Soils serve as a sink for excess rN where it can be stored for centuries if not millennia. The amount of rN in all the plants and animals on the planet is only about one-third of what is present in soils (U.S. Environmental Protection Agency, 2009).

Reactive nitrogen definition

Definition of available rN for the purposes of this book:

- rN that is not locked up in organic complexes (soils with carbon-to-nitrogen ratios generally of 10 to 12).
- rN not in other long-term storage.

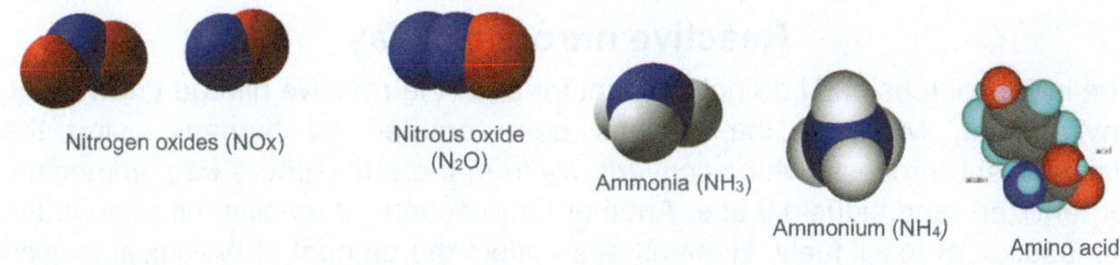

Figure 3.12 Various reactive forms of nitrogen.

Reactive nitrogen is limiting to biological systems because it is scarce and biological reactions stop when the available nitrogen runs out. For this reason, changes in rN availability have enormous environmental consequences.

So, if available reactive nitrogen is so scarce, how did it become available in the first place? When there is lightning, there is more than just rain and flashes of light. The electrical energy from the lightning splits nitrogen molecules, otherwise inert in the atmosphere, and releases single nitrogen atoms, which on their own are not stable. These atoms react with other constituents of air (principally oxygen) to form temporarily stable, but nevertheless reactive forms. These molecules fall to earth, generally bound to the water falling as rain. Once it becomes reactive, nitrogen can bond with other elements, such as hydrogen), and readily form a wide array of molecules (such as **ammonia** (NH_3) (Figure 3.12) and **nitrate** (NO_3^-) and be taken up by plants. It is estimated that some 5–8% of rN is produced this way each year. Nearly all the rest of naturally occurring reactive nitrogen is formed biologically by microorganisms in the soil or on the surface of water.

As opposed to N_2, reactive nitrogen changes its molecular form relatively easily. This presents one of the major considerations for environmental management. The three most important forms of rN that need to be remembered are as follows: nitrate , ammonia, and nitrous oxide (N_2O). Nitrate is the primary form that is taken up and utilized by plants. Ammonia results from the decomposition of natural organisms. Ammonia can be toxic to animals, fish, and humans depending upon its concentration, usually in water. A variant form of ammonia, ammonium (NH_4^+) (Figure 3.12), forms the core of amino acids, the building blocks of proteins that serve as biology's master builders. Nitrous oxide is a byproduct of the **denitrification** process, but not solely – the conversion of NO_3 to N_2 – whereby oxygen enters and disrupts an **anaerobic** system. The roles of rN as ammonia and nitrate are vital to life.

Reactive nitrogen today

The initial sources of rN do not account for all of the reactive nitrogen now in the environment. Much of the rN has been created by humans using the Haber-Bosch process, which converts N_2 from the atmosphere into ammonium for fertilizers and industrial use. Another large source of reactive nitrogen is the combustion of fossil fuels. Humans even affect the amount of biological fixation by their choice of growing nitrogen-fixing crops such as legumes. Soybeans are an example of legumes. Globally some 100 or more teragrams (a trillion grams, a billion kg, or one one-thousandth of a **gigaton**) of nitrogen is captured and thus made biologically reactive and useful through biological fixation (Bezdicek & Kennedy, 1998; Vitousek *et al.*, 2013). This is compared to roughly 3–10 Tg that comes from generation by lightning (Fields, 2004). The exact number of teragrams both from lightning and from biological fixation, a subject debated academically, is not important for us in this discussion. A rough estimate of the total amount generated naturally is important, however, when, as we will do in a later chapter, compare the number to how much reactive nitrogen is being introduced into our environment through human activities.

Reactive nitrogen can change its molecular form quickly in the biosphere. Ammonia under the right condition readily converts to nitrates and nitrates to nitrous oxide. Most common forms of reactive nitrogen travel easily and quickly through our **environment**. Nitrate is water-soluble and thus is transported with water as it moves across and through the landscape. Nitrous oxide is transported as a gas that reaches the upper atmosphere. Eventually over an extended period of time that involves many years, it converts back to N_2 or it reenters the cycles of transformation from one form of reactive nitrogen to the next.

Because the nitrogen atom, once freed of its partner atom, is so biologically reactive, natural processes capture it, making it again scarce. In a natural world, unaffected by human activities, rN is quickly taken up by living organisms and bound up in complexes that make it unavailable for further reactions. This capturing occurs by incorporating rN into longer molecular chains, such as proteins, or in dense organic complexes, such as soil humus. The rN is quickly recycled through the various media of our environment. The living world grabs it and holds onto it tightly in its various forms as quickly as it is freed up, whether as a small molecule such as nitrate or ammonia or in ts other manifestations in combination with other elements in the environment. Rarely, except with human intervention, is rN found free as nitrate or ammonia in rivers, streams, or the ocean.

Illustrated below are six examples of reactive forms of nitrogen commonly found in the environment (Figure 3.12). The mid-sized shaded or blue balls represent the nitrogen atom, the larger red ball the oxygen atom, the smaller white ball the hydrogen atom, and the darker gray the carbon atom. Under terrestrial conditions, these forms regularly transform from the one into the next through both **biotic** (and **abiotic** processes).

Plants are a good place to begin the examination of the cycling of reactive nitrogen. Plants take up and cycle rN from the soil, as do microorganisms. This nitrogen is used in the organisms' growth and biological functions. Some plants themselves bring their own nitrogen-fixing microorganisms with them. These plants, legumes such as clovers and soybeans, produce **root nodules** that are populated by microorganisms that fix atmospheric nitrogen into reactive nitrogen available to the plant. When plants die and their chemical constituents are released, primarily through **microbial** action, nitrogen is released, generally as nitrate or ammonia.

Different terms that apply to nitrogen

There are many different terms that all refer to nitrogen in one of its chemical forms. Nitrogen, the inert compound present in our atmosphere, has the scientific denomination of N_2, since it is in this non-reactive form, joined to its sister atom. [Major reactive forms that are created when N_2 is split are ammonium (NH_4) or ammonia (NH_3), one nitrogen atom and three hydrogen atoms; nitrate (NO_3), a combination with three atoms of oxygen; and nitrous oxide (N_2O), two nitrogen atoms that share one oxygen atom.] Nitrogen can also be referred to as a fertilizer, a nutrient, sometimes even protein. To make matters even more confusing, the amount of nitrogen (or mass) present can be measured in different ways. It depends upon whether one measures the mass of nitrogen in a given volume of air, water, or matter or as nitrogen as part of a molecule, such as nitrate.

Reactive nitrogen is a limiting factor for biological systems and for this reason ecosystems have developed in ways to maximize its reuse. Ecosystems vary in their efficiency in cycling nitrogen, but no system is perfectly efficient. Efficient systems capture the nitrogen within plant matter which can accumulate over time, storing nitrogen in soil organic matter. Not all reactive nitrogen, however, is destined for incorporation into living organisms. Less-efficient systems leak nitrogen, with some escaping as NO_3 and N_2O.

Microorganisms play a critical role in cycling organic matter and the rN it contains. The processing of organic matter by microorganisms depends upon the temperature and hydrologic conditions. These conditions determine the rate of decomposition and the molecular forms of carbon, nitrogen, and phosphorus generated by soil microorganisms. Where soils are not saturated and **aerobic** conditions exist, biological processes convert the reactive nitrogen into oxygenated forms such as nitrate. In wetlands with saturated soils, anaerobic conditions exist where denitrifying bacteria convert free nitrate (i.e., rN that has not been incorporated into new soil) back to nitrogen gas (denitrification). Wetlands, therefore, serve to complete the cycle of converting residual rN back to its inert form in the atmosphere. Nitrous oxide, laughing gas, is a biproduct of the denitrification process, which is no laughing matter. It is a potent greenhouse gas and stays in the atmosphere on average 114 years.

Nitrous oxide can leave the atmosphere by reacting with stratospheric ozone (Bond, 2009).

3.2.2.2 Nitrogen transport

Nitrogen can cycle through the environment in ways other than chemical transformation. Reactive nitrogen dissolves in water as nitrate. Once in water, it can be transported to streams, rivers, lakes, and estuaries. Although nitrate is valuable as a nutrient to aquatic organisms, excessive amounts of nitrate can cause conditions deleterious to higher organisms, such as fish or humans. These excesses are the result of additional rN, exceeding the amount of rN these systems can process, being introduced into natural biological systems. The excess rN upsets the balance between the organisms within these natural systems stimulating some species, often exotic aquatic plants, which benefit from additional rN at the expense of others. In aquatic systems, the species benefiting can dominate the native **flora** creating algae blooms. When these algae die, microbial populations spike as they feed on and decompose the algae, generally resulting in oxygen deprivations for other organisms in the system. Even with this proliferation of growth, all of the rN is not utilized, and the residual is transported downstream.

Clearly, transport by water takes nitrogen from one location with associated impacts to another location where it can have a very different effect on the environment. The different impacts are a result of changes in variables such as salinity, temperature, **pH**, oxygen level, and residence period, as well as the biological community present. Each of these variables affects both the positive and negative impacts of rN on the environment. One example would be from a river flowing into an estuary where by the same concentration in each can have very different ecological consequences. Nitrogen is also transported physically by air, animals, and in particular humans. Animal migration or livestock transport along with commerce in crops moves large quantities of organic matter, including nitrogen in the form of amino acids in protein. Agricultural production and food distribution result in nitrogen being transported as fertilizer, grain, livestock, and food. This transports nitrogen to areas where it can disrupt environmental systems in the new location that needs to process the additional nitrogen. Water, air, and other resources can be disrupted if the system cannot process ("fix" into longer term storage) the additional rN.

> ### What do we mean by 'saturating' a substrate or soil?
> The soil or a mineral substrate (something that is acted upon) is analogous to clothes hanger that can only accommodate six shirts. There are what scientists call binding sites, like the hook on the hanger, that can accept a molecule. Add more molecules than binding sites, and the excess goes into solution, just like my extra shirts wind up on the floor.

Ammonia is a form of rN primarily produced from the chemical breakdown of proteins in animal waste products but also from plant decomposition. It is most common in gaseous form and is also water-soluble. When it is released into the air as ammonia gas, it can be transported considerable distances under certain weather conditions. Wetlands – though generally a sink for rN – can also be a source. When wetland soils are disturbed either naturally through a geologic upheaval or by human alteration, nitrate and ammonia can be released, the former through water and the latter through air transport.

Reactive nitrogen ends its journey through the biosphere when it is either changed into a form that is no longer bioavailable, such as in soil complexes, in buried organic materials, or back into non-reactive N_2, nitrogen gas. In addition to wetlands, well-managed agricultural and forest soils also lock away reactive nitrogen and make it unavailable for further interaction with other components of the biosphere. Retention by these soils is less permanent than transformation back to N_2 or storage in deeply buried materials.

3.2.3 Phosphorus

At this point in our narrative, we need to introduce phosphorus (P) the other key nutrient that plays an outsized role in current resource management. Phosphorus, as has been mentioned earlier, is highly reactive and therefore does not occur in nature alone, but rather as phosphate (PO_4). It generally combines readily with oxygen atoms as a molecular complex that in turn can bind with other atoms, such as calcium, a mineral element. Unlike nitrogen that literally comes from the air, phosphorus as phosphate is generally found bound tightly to minerals in large rock formations. It is not normally water-soluble, but can become soluble when a substrate, such as soil, becomes saturated with phosphate leaving no more binding sites for the excess, which goes into solution with water. Phosphorus in the form of biologically available phosphate is scarce. Phosphate becomes available to plants when minerals containing phosphate are weathered and decompose into their component parts. What is not taken up by plants and microorganisms is quickly bound to soil or other substrates. Without human intervention, it is, like reactive nitrogen, scarce and biologically limiting.

In nature, large quantities of phosphate occur where animal waste has accumulated over large periods of time, such as deposits of **guano** from bird waste. The mineral **apatite**, a mined source of phosphate, was formed from marine deposits where sea organisms accumulated over many thousands if not millions of years. These organisms collected scarce phosphate made momentarily soluble through mineral **weathering** and then died, creating layers of sediment that over time became rock.

Together with reactive nitrogen and carbon, phosphorus plays a critical role in forming the building blocks of life. Like reactive nitrogen, too much phosphate poses an existential threat to higher organisms. Concentrations of phosphate in water bodies, something that normally does not occur in nature, represent a serious **pollution** problem for humans and animals with a similar effect as for excess rN.

Soils are a major sink for phosphorus. Soils retain a considerable stock of phosphorus but in a manner where only a small amount is biologically available. Phosphate is removed from the biosphere through the very slow process of mineral formation, a process whereby soils and sediment washed off the land into seas become deeply buried and become compressed into sedimentary rock. Mineral formation occurs at a temporal scale measured by millennia or eons.

How is phosphate transported from one location to another in natural systems? The most important natural modes of transport for phosphate are rivers and wind (Mackey & Payton, 2009; Science News, 2015). Weathering events that make phosphate available to plants can occur from thousands of miles away (Wind and Phosphate Box). As with nitrogen, agricultural production and food distribution result in large amounts of fertilizer, grain, livestock, and meat and the phosphorus they contain being transported between ecosystems. What this means is that phosphorus in biologically available form can be transported to locations where it is scarce or does not occur naturally.

3.2.3.1 Phosphorus pools and cycling

The important thing to bear in mind is that only a minute portion of phosphorus is biologically available. As explained in Figure 3.13 (the arrows in the figure above represents the major flows between pools). The vast majority of phosphorus lies in sedimentary rock. The mineral apatite is a major source of phosphorus. In the earth's crust, there is some 4×10^{15} metric tons. Compare this number to 2.8×10^9 metric tons of phosphorus in living biomass. Dead organic matter contains some 100 times more phosphorus, but the sum in living and dead biomass represents only about 0.01% of what is present in sediments and sedimentary rock (Canfield *et al.*,

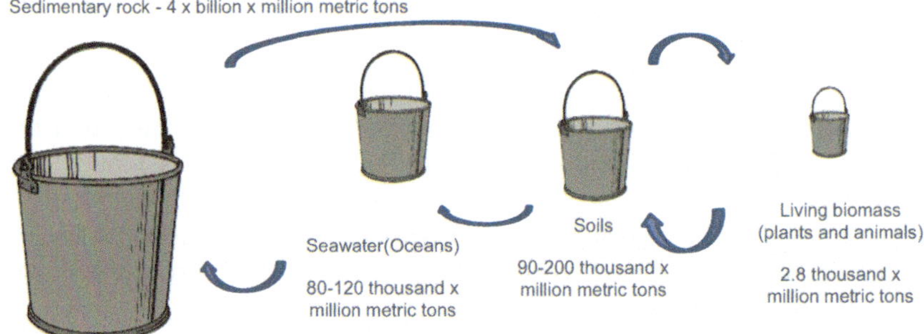

Figure 3.13 Phosphorus pools. Arrows indicate flows between pools.

2005). When viewed a bit differently, there is some $90\text{–}200 \times 10^3$ million metric tons (MMT) (90,000 to 200,000 MMT) P in the soils of the world. The bioavailable **phosphorus pool** is only some 1.805 to 3.0×10^3 million metric tons (1,805 MMT to 3,000 MMT) of phosphorus. Seawater holds some $80\text{–}120 \times 10^3$ MMT (80,000 to 120,000 MMT) of phosphorus (Figure 3.13).

Wind and phosphate

Wind transport of phosphorus – for thousands of years, phosphorus in dust blown by the wind has served to fertilize areas thousands of miles away. A prime example is the seasonal winds of the Sahara, the so-called Harmattan wind that carries phosphate-rich dust from the deserts of North Africa as far as the eastern shores of North and South America. These winds can carry some 400–700 million tons of dust in a year (Adetunji, 2020).

Although weathering makes new phosphorus biologically available, the descent of detritus to the bottom of water bodies completes the cycle. This phosphate has been transported by rivers and wind to oceans and lakes. Most of it combines with the element calcium to form calcium phosphate. The uplifting of the land exposing the sediments to the weather renewing the process takes eons. In contrast, the process that occurs on land with the transfer of phosphorus from the soil to plants, then to animals and humans, and back to the soil, generally mediated by water, proceeds over weeks and months (Liu & Chen, 2008).

3.2.3.2 Summary

Our major ingredients of life, carbon, nitrogen, and phosphorus, occur in many forms, often changing from one into another and between different states, solid, liquid, and gas. But, the mass, that is the number of atoms, is conserved. Just as in our play with LEGO™ pieces, the number of pieces at the beginning of our day when we constructed our imaginary world of people and things remains the same at the end of the day after all has been deconstructed and put away for play on another day.

Chapter 4

Soils, or how dirt plays a part in the precarious balancing act of cycles

The Nation that destroys its soil destroys itself.

— Franklin Delano Roosevelt, 1937

Soils are the Rodney Dangerfield of natural resources: 'They get no respect'. We walk all over them and call them dirt. But they are essential for life as we know it, providing nourishment for plants, filtering and storing water, and serving as a habitat for insects and microorganisms that are critical for cycling carbon and nutrients. Understanding the services that soils provide is essential for effectively managing the environment and protecting and conserving natural resources. From the perspective of humans, soils are vital, for we would have no **agriculture**, no food, and no civilization as we know it without this essential resource. Soils interact with our atmosphere through the hydrologic cycle by determining the amount of evapotranspiration which in turn can influence the local climate (Abramopoulos *et al.*, 1988; Miller and White, 1998; US Global Change Research Program, 2014). They even affect the temperature we experience. Soils play the same role for the earth as our skin does for us, protecting the internal us from the elements (Suszkiw, 2017). Soils also keep us safe by preventing nasty stuff, like toxic metals, from contaminating the biological world in which we live. Last, but not least, soils, as we will show, affect how water flows and thereby the transport of carbon, nitrogen, phosphorus, and other materials.

© IWA Publishing 2021. A Guide to Understanding the Fundamental Principles of Environmental Management. It Ain't Magic: Everything Goes Somewhere
Authors: Andy Manale and Skip Hyberg
doi: 10.2166/9781789060997_0057

The role of soil as a regulating medium is often overlooked, but it is essential. Although nature constantly changes, it seemingly stays the same. The seasons come and go, temperatures rise and fall, storms rage, and droughts happen. Dynamic processes cycle with the physical, non-living world dictating conditions and flows in the biophysical world. And yet with all the dynamism we observe, the world stays within familiar boundaries – balanced, that is, in harmony so to speak. We take our current comfortable conditions for granted. However, within these biogeochemical states, the key ingredients of life – carbon, nitrogen, and phosphorus – cycle. They change molecular form and state (gas to liquid to solid). Over the expanse of time, much of what is reactive has gone into long-term storage in the earth's crust, deep ocean waters, sediments, or, importantly, in soils. The interplay of the cycling of the reactive components with the physical states establishes limiting conditions for biological organisms. These components also interact with physical or environmental factors, such as temperature and water. A change to this balance has repercussions on our physical conditions, the quality of our environment, and the natural world we value.

In this chapter, we discuss how natural systems – with soils at their core – regulate the availability of reactive forms of carbon, nitrogen, and phosphorus to make them scarce within the environment. In later chapters, we discuss why this is important and how changing the availability of these reactive forms jeopardizes higher, more complex life forms with which we are most familiar – animals and plants, and humans – and thus the quality of human life.

4.1 WHAT ARE SOILS?

Soils are the nexus between air and water resources, plants and animals and the habitats in which they exist, and the physical world of reactive (bioavailable) versus nonreactive (inert) chemical components of our biosphere. Soils provide the conditions necessary for carbon, nitrogen, phosphorus, and other nutrients to be taken up by plants, decomposed by insects and microorganisms, and transformed within their cycles. These processes help determine environmental and resource quality in our contemporary world and will be discussed later, but for now let us answer the question, 'What is soil?'.

The simplest explanation is that soils are the skin of the earth, the layer of material that interacts with air, water, and most terrestrial vegetation. However, a more technical description is useful for understanding how they function within the natural cycles we are examining. So, let us use the definition – a soil is a dynamic **matrix** of mineral particles, air, water, organic matter, and living organisms. The proportions of the **inorganic** components, that is, non-carbon compounds, determine the properties a specific soil will have. Two of these components, water and air, can be very dynamic, changing with the weather and other processes. The mineral component is generally very stable, with changes taking place on a geologic time frame. The organic component changes with changes in management, land use, and vegetative cover. Soils play a role in capturing essential nutrients from our environment, filtering what enters the terrestrial realm, regulating how much water may infiltrate and thus become groundwater or be available for runoff, and protecting earth's components within and under soil from dispersal (see http.www.soils.org website for greater explanation of what soils are, how they form, their different types, and their importance).

4.2 CYCLING THAT OCCURS ON A HUMAN SCALE

For resource management, it is the bioavailable, reactive portion of the mass of carbon, nitrogen, and phosphorus that is our focus. These bioavailable forms are maintained at a, more or less, constant proportionality (ratio to each within the soil and ocean pools). In other words, over time, their proportions tend toward an equilibrium or balance (Cleveland and Liptzin, 2007; Fang *et al.*, 2019; Griffiths *et al.*, 2012; Tipping *et al.*, 2016) whereby if one mass changes, the others will adjust accordingly. This applies to soils in temperate climates as well as in the arctic or equatorial regions and oceans throughout the globe. [Because this book is focused on the management of terrestrial systems, we mention the importance of ocean processes that drive the balancing of carbon, nitrogen, and phosphorus, the so-called Redfield Ratio (Gruber and Deutsch, 2014), but elaborate the concept using soils.] By serving at the core of cycling processes, they connect with longer-term storage pools whereby excess portions are locked away, no longer available for the biological processes we observe. The biological processes in these pools, by balancing the ratios among these elements, play a core role in ecosystem functioning.

Because of the importance of soils to environmental and resource management, we repeat points made earlier. Soils are composed of organic matter (decomposed plants and animals), weathered **bedrock** minerals, air, and water. In them, the chemical elements of life (carbon, nitrogen, and phosphorus) have been incorporated into biochemical complexes that scarcely resemble the microorganisms, plants, and animals from which they are derived.

You can imagine the components of the water, carbon, nitrogen, and **phosphorus cycles** as diverse gears that mesh with each other (Figure 4.1). The gears turn, each at their own speed – some fast, some slow, and others at rates almost imperceptible to us. Changing the size of the gear or speed at which any one of them turns affects the others, though the consequences may not be observed for a long time. The relationship among **bioavailable** carbon, reactive nitrogen, and bioavailable phosphorus is analogous to these gears. As we point out and emphasize in the chapters of this book that follow, human actions that interfere with the meshing of these cycles have caused consequences that can take years to play out. Because this book is about managing natural resources and the environment – a positive message – we lead the reader to the questions and the tools (practices

Figure 4.1 The nitrogen, carbon, phosphorus, and water cycles are closely connected to one another, but the speed at which the cycles respond to changes can differ. Image source: Wikicommons.

and policies) for restoring balance, if not mitigating the adverse consequences associated with imbalances, in later chapters.

4.3 SOILS: WHERE THE CARBON, NITROGEN, AND PHOSPHORUS AND WATER CYCLES MEET

At the heart of these dynamic interactions lie soils. We repeat this point made above because of its importance. It is in soils that the biochemical reactions take place that enable carbon and bioavailable nitrogen and phosphorus to be stored. It is here that relatively constant ratios between these critical elements (see box on Carbon, nitrogen, phosphorus soil ratios) are maintained. Soils are the foundation upon which the ecosystems depend. And ecosystems are what we depend on for our existence.

Soils serve as the third largest sink for carbon, after earth's crust and the oceans. They are a major sink for carbon, rN, and phosphorus, but are variable in how much they store. Soils in environments with little water contain much less carbon than grassland, **prairie**, or forest soils that are often converted to agriculture. For example, US temperate grasslands, which have a generally moderate climate, have on average 189.1 Mg per hectare of total carbon per square kilometer (Lal *et al.* (eds.), 2017 and 2018b). In contrast, desert scrub averages 57.8 Mg per hectare, while swamp and marsh have on average 725 Mg per hectare (Lal *et al.* (eds.), 2017 and 2018b). Wetland soils are effective in storing plant material and making it unavailable for further reaction within the biosphere. They take carbon dioxide out of the air and hold it in the ground (carbon sequestration). Organic soils associated with wetlands sequester much carbon and can have organic depths of dozens if not hundreds of feet (Sazonova *et al.*, 2004). Wetlands are so effective at this that over periods measuring tens of thousands of years, these soils could actually lead to significant **reduction** of atmospheric CO_2 levels and thus a global cooling if no disturbance occurred to release the stored carbon (Lal *et al.*, 1998a, b; Nahlik and Fennessy, 2016). Some have suggested that the process of land storage of a massive amount of carbon contributed to past ice ages (Rayner *et al.*, 2016). These soils can accumulate massive amounts of carbon that can become, after burial by sediments and over hundreds of thousands and millions of years, petroleum and coal deposits (National Geographic, 2020).

Intrinsically intertwined with soils is the earth's vegetation. Trees contain carbon, and forests due to their extent contain large amounts of carbon. Most, if not all, of us are familiar with the role of the Amazon basin in sequestering carbon. Additionally, large portions of Siberia, Africa, North America, and Asia are covered by forests which store and continue to accumulate carbon. In examining carbon balances and changes over time, we should recognize that most of Europe and the eastern United States was forested hundreds of years ago (Belmaker, 2018; Oswald and Smith, 2014; Roberts *et al.*, 2018). Old German tales and historical records from the colonization of the Americas tell of lands where forests once dominated are now agrarian landscapes. The conversion of these landscapes from forests to other uses released carbon that had long been stored as wood and, just as importantly, within the soils. These conversions can be reversed. In the Eastern United States, a good deal of land that was converted from forests to cropland has reverted back to forests over the past century. Because forests change their appearance gradually, many people assume that the forests that they currently enjoy have been present for essentially forever. They do not realize that most of the eastern forests have actually been reestablished as more fertile cropland was discovered, eastern cropland abandoned, and new energy sources adopted that reduced the use of wood as fuel (trees provided much of the fuel that powered the eighteenth and nineteenth centuries). This **reforestation** is an example where ancient forms of organic matter (petroleum) have replaced geologically newer forms of organic matter (trees) as the preferred fuel and have ironically led to these second growth forests. As these forests grow, they will sequester carbon in the stems and branches of the trees, and within the forest soils (Boyer and Groffman, 1996; Heath *et al.*, 2003).

Where drier climates marked the natural transition from forests to grasslands, grass species thrived. The detritus from the annual dormancy and renewal of grassland vegetation contributes to the formation of carbon rich soils. Although these grasslands typically store less carbon than forestlands annually, over millennia grassland soils can store large amounts of carbon, making them highly productive for agricultural production (Lal *et al.*, 1998a, b; Potter *et al.*, 1999). It has been estimated that these soils contained approximately twice the amount of carbon as the same soils that have been converted to cropland (Donigian *et al.*, 1994). These grasslands and the croplands that have replaced them can still serve as important sinks for carbon if these lands are managed in a manner that incorporates plant material into the soil (Potter *et al.*, 1999; Ruddiman, 2018).

Wetlands form in areas where soils are saturated and water fills the soil, creating **anaerobic** conditions. In the United States circa 1780, there were some 81 million hectares (200 million acres) of wetlands sequestering vast amounts of carbon, nitrogen, and phosphorus (Dahl, 1990; Dahl and Allord, 1997). These wetlands also are critical to the nitrogen cycle because the anaerobic conditions within their soils combined with a high carbon content provide the conditions needed for denitrification, completing the **nitrogen cycle** by returning nitrogen back to the atmosphere.

Importance of carbon, nitrogen, phosphorus ratios in soils

What is the importance of the observation that there are relatively constant ratios among carbon, nitrogen, and phosphorus in the organic portion of soils to natural resource management? How should we interpret the observation that the various key cycles in the biosphere are linked to each other in a constant relationship? How can this knowledge help us make better management decisions? The graphics below help to explain.

The carbon–nitrogen–phosphorus ratio is on average 108::8::1. Adding additional organic material to soil will increase the mass of carbon, nitrogen, and phosphorus in the 108::8::1 ratio. Here, we show the added organic material doubling the mass of these elements within the soil. This would take years or decades to be completely integrated into the soil. If added to the soil in a different ratio, then the excess amount of the more abundant element(s) will be lost to the environment. The largest slice represents carbon, the next largest nitrogen, and the thinnest slice phosphorus in the above-mentioned ratios.

Imagine that Figure 4.2 represents the soil organic matter in your garden. The slices of the pie represent the ratio of the elemental components. The average relationship for the amount of carbon, nitrogen, and phosphorus in world soils in temperate regions is a ratio of 108 carbon to 8 nitrogen and 1 phosphorus, measured by mass, or 108::8::1 (Tipping *et al.*, 2016). The total mass at the end of the season will depend upon temperature and rainfall. The ratios among these elements are not absolute; but they vary within a fairly tight range. The proportions vary with soil depth region, climate, and other factors. For example, the ratio of nitrogen to phosphorus declines slightly with increasing soil depth. And tropical soils tend to have less available phosphorus, resulting in a slightly higher carbon-to-phosphorus ratio. However, for temperate climates, those we use for most examples, the 108::8::1 ratio holds. The excess of any one of the elements moves to another pool (see Chapter 3) or is lost to air and water

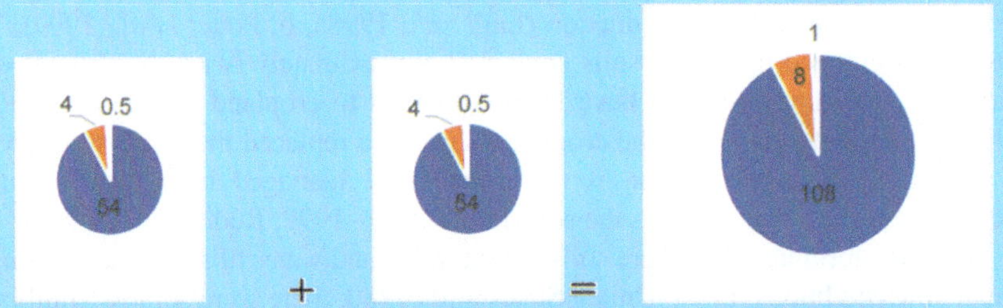

Figure 4.2 Proportions of carbon to nitrogen to phosphorus in healthy soils.

where it can be transported offsite. To the extent that carbon, nitrogen, and phosphorus move to a longer-term storage pool, they are removed from bioavailability.

In this example, let us now add more carbon, perhaps, we add straw to our garden. Will the slice become larger in accordance with the additional carbon? The slice can only expand in size if carbon, nitrogen, and phosphorus are added in the proper ratio. If our garden contains 54 units of carbon, 4 of nitrogen, and 0.5 unit of phosphorus and we add 54 units of carbon along with four units of nitrogen and half a unit of phosphorus, we double the size of the slice (Figure 4.2). If we add 54 units of carbon along with 8 units of nitrogen and 1 unit of phosphorus, the pie still doubles, but 4 units of nitrogen and a half unit of phosphorus are released into the environment.

Now imagine what happens if we grow a crop in our garden that removes half the nitrogen in the soil. Our slice now contracts to half its former mass. Stated differently we have removed 4 units of nitrogen from our original system. Half the carbon and half the phosphorus will be lost from the organic complex to restore the original balance. The total organic mass reduces by half. Soil organic mass can also be reduced by burning away, that is oxidizing the carbon, thereby releasing the nitrogen and phosphorus.

As we explain in Chapter 5, farmers have known for thousands of years that soils have to be either rested or fed the equivalent of the nutrients that are lost when crops are harvested or the quality (or total mass) of the organic component of soil will be lost. Over time, if the soil is shallow, the land becomes unproductive. Conversely, soil can be replenished by adding new carbon, nitrogen, and phosphorus at appropriate ratios.

4.4 THE UNDERAPPRECIATED SERVICES OF SOILS

Soils are a natural resource that provides benefits (i.e., services) for the public good. [See Appendix D: Ecosystem Services.] For this reason, the European Union provides soils with legal protection (European Union, 2007). The overarching framework directive of the European Union identifies pressures and impacts on soil quality and their protection. This includes short- and longer-term objectives and strategies to achieve them. In the United States, there is no comprehensive federal strategy dealing with all soils but rather targeted actions to reduce the impacts of poorly managed (particularly highly erosive) soils, in agricultural use (Natural Resources Conservation Service, 2020a). There is little reliable data on the health or condition of soils for most of the developed world, particularly those that have been under intensive use for hundreds of years. This lack of reliable information inhibits the development of policies to better manage soil resources (Cerri *et al.*, 2000).

Although it is important to understand how past use has affected this essential resource, you do not need these data to appreciate the value of soils, the benefits from sound soil management, and the consequences of their neglect. You only need to understand a few basic facts about soils and to be aware of their connection with other two essential resources – air and water.

Soils regulate surface and subsurface processes by controlling and limiting (regulating) the availability of the carbon, nitrogen, and phosphorus, thereby helping to provide balance within ecosystems. Such constraints are necessary when too much of something will disrupt ecosystems, with the imbalances potentially cascading through the environment to harm other systems. Balance permits biological systems to grow and to thrive. [Note that balance, as we are using it here, does not mean a static system with no change. We are using balance to refer to that state when systems are in equilibrium (more correctly, one would say **steady state** but that is a distinction for specialists in the field.] By providing these regulatory functions, soil with its position within the landscape greatly influences surface and subsurface biology and hydrology.

What makes up soils and how are they formed? There are over 18,000 soil series recognized in the United States, and several soil series can often be found in the same field. This variability can make managing the soil resource challenging (National Soil Survey Center, 1995). The quality of soils differs tremendously, starting with the material that comprises them. The material can be minerals dissolved or weathered from bedrock or sediment deposited over time by wind or water. This material is coupled with organic material from vegetation or deposited by macroorganisms and digested and processed by microorganisms. The formation of the organic complex is the gradual bit-by-bit deposition and decomposition of dead bugs, plants, and micro- and **macroorganisms**. Soil formation takes thousands, if not tens of thousands of years (Millennium Ecosystem Assessment, 2003). But if we take into consideration how long it took for the chemical ingredients of soils to develop and to achieve the balance among reactive and nonreactive, biologically available and inert components, the timespan extends into hundreds of thousands if not millions of years (Birkeland *et al.*, 2003; Burns et al,, 2003; Soil Science Society of America, 2013). How fast and how much soil is formed depends not only on the minerals that are weathered and the organic material deposited, but also on climatic conditions, particularly temperature and water availability. As we repeatedly witness, soils, water, air, and climate are interconnected in which the cycles (gears) in this complex machine turn at different speeds and thus proceed over differing timeframes.

Dig into undisturbed earth with a shovel so as to make a cross-section and you will find layers (or **horizons**), of different colors, textures, widths, and even smells. These layers reflect different periods of soil formation as well as the accumulated effects of deposition of organic materials, water, **microbial** processes, **erosion**, **tillage**, and other influences over time. These layers affect water infiltration, retention, and lateral movement. In cases when the soils have been disturbed by plowing or development, the upper layers are mixed.

What services do soils provide us? Why are these services so important to humans and to the ecosystems upon which we depend? We need soils to grow forests, grasses, and the crops that provide food, fiber, and energy. Soils help regulate the flow of water through our landscapes and sub-surfaces, storing and filtering water. The organic matter in soils acts like a sponge soaking up and holding water in place. This sponginess can make the difference between a flood event or simply high-water flow. By taking up water in pores within its matrix, soils store water helping to counteract periods without precipitation. Take, for example, peat moss that gardeners put in potting mix. That addition of extra organic matter, carbon, keeps the water in place. In addition, it allows the soils to breathe – the exchange of carbon dioxide with oxygen that is associated with plant growth. Also, by filtering what goes into the ground, soils purify water destined for ground water and aquifers. Soils buffer water, reducing its acidity. This is important for keeping toxic substances from contaminating the biosphere and maintaining a balance between the mineral world and the biosphere. In summation, soils by holding, releasing, and purifying water are an important regulator in restoring balance to the land when events aboveground, such as floods, droughts, or perhaps climate change wreak havoc on exposed life.

Experiment 4.1: Plant beans in two pots, one containing quality topsoil and another with subsoil or sand that is largely bereft of its organic component. Compare the growth of the beans growing in these pots. You will observe that the beans planted in the subsoil will grow slower and likely exhibit or signs of ill health such as yellowing and wilting.

Remember that soils are not inert media, such as cement or plastic mailing peanuts, but are living systems that we observe interacting with **flora** and fauna. Soil microorganisms and soil insects living in soils play a critical role in processing nutrients and carbon. Carbon, nitrogen, and phosphorus cycle through soils as they play important roles in the functioning of our biosphere. Soils preform this by providing a habitat for the plethora of small vertebrates, invertebrates, and microorganisms that function as recyclers. Remember that rN and bioavailable phosphorus are naturally scarce, thus highly valuable. Organisms evolved to use, decompose, and thereby recycle most if not all of these nutrients from deceased organisms by breaking down even the largest creature into its component parts. What was once waste is converted into temporary storage as part of soil until the above ground plants draw upon these nutrients. Thus, the health of surface plants and animals is tied with that of the soils below ground. Healthy soil, rich in microorganisms with a balance of carbon, nitrogen, and phosphorus, enables vibrant plant growth.

In addition, soils serve as a depository for **genetic diversity**. If you have ever tended a garden, you know that seeds, especially seeds that are not particularly wanted in our garden soil such as weeds, are stored in the soil. Some of these can sprout decades after they were deposited (Kalamees *et al.*, 2012). And there is more and more scientific evidence, that soil microorganisms share genetic material with the macroorganisms, like plants, that we use and consume. The DNA stored in microorganisms may serve to replenish the biological diversity of systems depleted of biological species (Neal *et al.*, 2020).

Living legacy

These other components of soil, which are the legacy of thousands if not millions of years of natural processes preparing the groundwork, literally, for all the higher organisms that make up our biosphere. Variations in soil help determine the subtle differences that make our planet so diverse and interesting.

Remember that hundreds of millions of years passed before life on earth could establish itself. This required the separating out and sequestering chemical elements anathema to life, such as lead, cadmium, mercury, etc., in mineral formations and then processing mineral complexes, such as silica minerals, into new complexes that would then lend themselves to life. Remember again, that the organic world of carbon lives in a symbiotic relationship with the inorganic world, largely of silica. In early earth, silica minerals served as catalysts for organic processes. Nature built these chemical complexes of non-organic chemicals over a vast expanse of time that then enabled the evolution of our complex organic world that we take for granted today. Regard our terrestrial world representing one third of the surface of the earth, the first step in the biological conquest of land may well have been the creation of soil. The idea that earth is a living planet involving both its organic and inorganic parts is indeed not just a poetic statement but a scientific truth.

4.5 SOIL PROPERTIES

Soil, like the world above it, abounds with plant and animal life. As everywhere in our living world, the key ingredients for the recipe are carbon, nitrogen, phosphorus, and water. Clearly soil and its properties are not the same everywhere. We all know this because we have experienced even in our backyards, how some soils pool water while other soils drain water as fast as we apply it. And if you think about it for a bit you probably have observed that some soils are red while other soils are as black as tar, that some soils produce plants that grow as tall as trees, and other soils grow nothing but stunted weeds. Why then are there such differences between soils? Partly it is because of source material, partly because of how and where a soil is formed, and partly because of what happened to the soil after it was formed. A soil formed in a dry landscape with extremes of heat and cold will differ significantly from one formed from sediments deposited in the ocean. This in turn will differ from one formed in the depression of base rock from material deposited by water, such as snow or ice melt from glaciers. Each of these soils has different base materials and climates in which they formed. And to throw in one other factor, the level of the water table will play a role in determining the biotic component in the soil and can either inhibit or nurture plant growth and chemical processes.

What are the important soil properties that make them a keystone for environmental processes? **Soil texture**, perhaps the most obvious of soil properties, is the proportion of different sized mineral materials (clay, sand, and silt). The size of soil particles affects the infiltration of water. As was discussed earlier, the size of a particle determines the size of the pores between particles, which in turn determines how rapidly water moves through the soil. The larger the pore is, the more rapid the downward movement of water. Soil texture establishes the capacity of the soil to retain moisture, as well as the nutrients present in water. Finally, soil texture, along with soil carbon, governs the stability (reflected in its structure as explained later) of soil **aggregates** (soil particles or clumps), and the propensity of a soil to erode.

This soil structure – its clumpiness – depends upon the site-specific conditions under which the soil formed. Soil structure is the grouping or arrangement of soil particles (Brady, 1974). Soil aggregates (cohesive groups of particles) are important for water infiltration into soils. The more that soil particles are aggregated, the larger the pores within the soil. To verify that increasing pore size increases water infiltration, conduct the following experiment.

Experiment 4.2: Remember our experiment in Chapter 2 where we poured water on clay and sand. You could modify that experiment and pore a gallon or liter of water on the ground and watch how much of that water flows downhill. Do this on sandy and clay soils and observe how much stays on the surface and how much infiltrates. It depends upon the soil's properties, particularly pore size which is a function of soil texture and soil structure.

The stability of these cohesive soil particles is an important characteristic of the soil structure. If the aggregates break down easily, then the soil structure will not be maintained. The colloidal components of soil are the finer fractions of the soil that are the most chemically active. These materials when placed in water stay in suspension in solution or settles very slowly if at all. Soil **colloids** play an important role in both binding soil particles together into aggregates and stabilizing them.

We need to take a little side trip here to talk about **colloidal particles**. Colloidal particles in soils are either clay or humus. The particles are extremely small (less than 1.0 **micron** in size), have a large surface area per unit weight, and have surface charges that attract ions (atoms with an electric charge such as calcium (Ca^+) or magnesium (Mg^{++}), or charged molecules like ammonium (NH_4^+) or phosphate (PO_4^{3-})) and water (Goldberg et al., 2012). For these reasons, they influence soil structure and soil chemical properties. The ability of these particles to attract and hold ions make them critical intermediaries in plant – nutrient interactions because the surface charges of colloidal particles function as the site of the carbon, nitrogen, and phosphorus chemical reactions and exchange. And, they are important for soil structure. 'Because of their surface charges, colloidal particles also act as a 'contact bridge' between larger particles, helping them to maintain stable granular structure. The surface charges on colloidal particles make them the center of soil chemical activity and nutrient exchange.' (Brady, 1974) What does that mean? It means these particles serve as the stage for some of the most important activity occurring in soils, including soil aggregation, nutrient retention and exchange, and water storage. The colloidal particles are where the carbon, nitrogen, and phosphorus action takes place and the gears of the cycles mesh.

Experiment 4.3: To observe colloidal particles put a tablespoon of the top layer of deep black soil into a glass of water. See what happens. Wait a day, the water in the glass will not be clear. It is the colloidal particles that are discoloring the water.

Soils with more colloidal material (clay and organic matter) have more stable clumps of soil. Because these colloidal particles are small enough to interact with other particles at the molecular level and because larger particles do not form these bonds, they can develop **chemical bonds** with other soil components to form larger clumps. These clumps (or aggregates) with high components of clay particles are generally more stable than aggregates in soils high in sand (An *et al.*, 2010; Brady, 1974; Skidmore and Layton, 1992). Soils with the right amounts of colloidal particles are good for growing things, including crops. They can hold and store more water and the nutrients in water. And they enable roots to grow, benefiting plant development and health.

Soil organic matter, which contains much of the biologically available carbon, nitrogen, and phosphorus, has been discussed numerous times before. It generally comprises 0.5–10% of a soil, but can approach zero in highly eroded and arid soils (Brady, 1974; Haynes, 2005). It can be much higher in wetland soils. Organic matter is made up of three components: small (fresh) plant residues and insects and **microorganisms**, decomposing (active) organic matter, and stable organic matter.

The portion of the organic matter of soils that is far less easy to breakdown (**labile**) is called **humus**. It does not turn over year to year (Natural Resource Conservation Service, 2020c). It is formed through the decomposition of larger organic material and can be broken down or destroyed more rapidly than clay (Brady, 1974). Humus is directly involved in developing and maintaining soil aggregates. Humus is less dense, is made up of generally smaller particles, and has a substantially higher number of binding sites for charged molecules like rN and P than clay. Its substantial colloidal component provides the glue that helps bind together soil particles, contributing to aggregate stability. Because humus slowly decomposes from constant interaction with the environment, continued deposition and decomposition of organic material is important to maintain the humus that is in the soil. As will be explained in the next section, its small size, lighter weight, and position close to the soil surface cause humus to be more easily removed through erosion.

4.6 SOIL EROSION

Why is erosion, the displacement and movement of soil particles, an important consideration in managing natural resources? Eroding soils pollute water and the air because they transport sediment, nutrients, and chemical from the land into downstream waterbodies and the air. Soil erosion diminishes ecosystem health along with agricultural productivity. These problems are discussed in next few chapters. For now, we will focus on the factors that make soil susceptible to erosion.

The properties of a soil are fundamental to its susceptibility to erosion. Here, we provide a basic understanding of soil erosion and the soil properties that influence its propensity to erode. However, soils are dynamic with complex interactions among different soil properties. Assessing the erosiveness of soil using broad statements can be misleading. Understanding how the characteristics at a particular site affect soil erosion requires the assistance of a trained professional. However, one can say with confidence that soil texture, organic content, and soil structure each play an important role in the capacity of water to infiltrate into a soil, for soil to hold water, and for a soil to withstand erosive forces. Knowing the soil properties is only half the battle, one also needs to know the forces that soils are exposed to. This exposure is a function of climate which determines precipitation and wind energy. Knowing the soil properties and the characteristics of associated climate at the location of the soil tells us how susceptible the soil is to erosion. The characteristics of the landscape (particularly slope), land use and conservation practices, which determine the exposure of the soil to wind, precipitation, and surface runoff are important to determining how much erosion will occur. These will be discussed in later chapters.

Soil texture, the proportion of sand, silt, and clay in soil, and soil structure are important determinants of soil erodibility. They affect the amount of energy needed to move individual soil particles. Larger sand particles and clumps are heavier and less likely to stay in suspension. They are therefore less easy to transport. The converse is true for smaller particles – those most important for soil fertility– that are removed more easily because of their smaller size. Extremely fine particles can remain in suspension for long periods and settle only in still water (Brady, 1974; Gavrilescu, 2014). Soil organic matter, because of its role as a glue holding particles and clumps together, also affects the vulnerability of soil to erosion. A soil with stable aggregates and good soil structure does not erode easily.

Erosion is a function of the energy to which the soils are exposed from the water running and wind blowing over the surface of the land. In addition to larger sized soil particles and more stable aggregates being harder to move, they also create larger sized pores allowing more water to infiltrate leading to less surface runoff.

Less surface runoff, in turn, lowers the amount of energy available to move soil particles, helping to reduce erosion. The analogy of pouring water on bowling balls versus sand in Experiment 2.2 is a good example. Water will flow much faster through the bowling balls than the sand, and it is clear that it will take much faster moving water to move a bowling ball than a sand particle. Because greater infiltration reduces available (kinetic) energy soils are exposed to from runoff, the larger pore space reinforces the resistance of larger particles to erosion. We discuss this more when we examine the effects of human activity.

Human development, particularly in the last several hundred years, has replaced large amounts of native vegetation with agricultural crops, livestock pasture, and urban/suburban land. In most cases, the loss of native vegetation has led to less protective cover for soils. Less cover has meant more soil erosion. We will focus on agricultural lands because they comprise a large proportion of the landscape. Moreover, how the conversion of native vegetation to agricultural use affects **soil exposure** to erosive forces has been carefully examined in the scientific literature.

In examining the impact of the conversion of native vegetation, we do not wish to suggest that we are advocating for a return to a native pre-Columbian landscape. We are, in fact, not advocating any change from one type of land cover to another. We provide examples so that the reader can understand the changes that have taken place or that could occur and their effects on environmental management. They establish a measure for how much vegetative protection could be obtained. The appropriate baseline for determining a course of action is the current state of vegetative cover. This should then lead to an assessment of proposed options that could alter that state.

As stated previously, most areas of natural vegetation suitable for agriculture have been converted to cropland and pasture. Even in areas not directly converted to agricultural use or human space, livestock grazing has altered rangelands where native fauna had previously consumed native plants. Over time, the intensity of agricultural activity has increased. Even forests have not been spared. Much of North American forestland has been harvested at least once. This changed the species composition and tree density (other factors such as disease and invasive species have also altered the species composition in North American forests). Collectively, these changes have led to greater exposure of soils to erosive forces. The importance of agricultural productivity and the observation of environmental problems from erosion have compelled researchers to examine how they relate to each other. These studies have supported the development of conservation methods that address erosion. These methods are discussed in Chapter 6.

Vegetative cover includes all vegetation living and dead that protects the soil from wind or the force of moving water. This includes the overstory, understory, and ground cover (including plants, leaves and other residue) on the soil surface.

Plants also protect the soil by establishing roots that slow **overland flow** and bind with soils (Kramer and Weaver, 1936; Kramer & Weaver, 2015, 2015). The effectiveness of vegetative covers is a function of the portion of the soil they cover, their presence and the coverage provided throughout the year, the resiliency of the cover over time, and their ability to dissipate energy.

It is easy to see that removing vegetation from the soil surface exposes the soils to the forces of wind, falling rain, and overland surface flow. However, soil exposure to erosive forces can also be increased by changing the vegetative cover should the new cover expose the soil to greater erosive forces than the previous vegetation. Agricultural examples of this would be the growing of corn in place of hay or soybeans in place of corn. In both of these cases, the vegetation replacing the initial crop provides less soil protection than its predecessor.

When soils are exposed to erosion is important for understanding the function of vegetative cover. In climates where wind or rainfall amount and intensity are distributed evenly throughout the year, cover is equally important year-round. If the rainfall is seasonal, particularly if it is dominated by heavy storm events (e.g., monsoons), then the vegetative cover during the rainy season is more critical for protecting against erosion. In this way, the rainfall distribution plays an important role in determining whether a particular vegetative cover effectively protects the soil. Similarly, for soils protected by snow or frozen ground for portions of the year, vegetative cover is more important in those seasons when snow or frozen ground does not protect the land.

The protection afforded by different vegetative covers varies with the density of the growth and the crown closure. At the two extremes are a closely grown, all season perennial grasses that provide a very high level of soil protection and a bare soil providing none. Crops that are closely grown, such as wheat, oats, and hay, provide greater protection during the growing season and after harvest than row crops such as corn, soybeans, or cotton that leave the soil exposed during portions of the year. As **crop residue** after harvest increases, so does soil protection. Wheat, oats, and hay provide greater post-harvest soil protection than do row crops because harvesting leaves their stems and roots standing and intact, while harvesting row crops knocks over stalks and stems.

Forest land with intact understory and ground cover provides good soil protection, but generally somewhat less than is provided by perennial grass. The understory and ground cover beneath the tree canopy are important for protecting soils both to slow overland flow and to protect the soil from water dripping from the canopy. This is because water dripping from heights over 8 meters will reach terminal velocity by the time it reaches the ground, thereby having the same impact as rain. If there were no understory or groundcover, the canopy would not protect soils from drip erosion (Satterlund, 1972). Note that there would be a reduction in the amount of water reaching the ground due to evaporation from the leaf surface.

4.7 THE SOIL BIOME – MICROORGANISMS MAKE THE CYCLE GO

The soil microorganisms, fungi, and other life are what make soils a keystone in the cycles that we have discussed throughout the first three chapters. The biochemical process that converts atmospheric nitrogen into rN is conducted in the soil (and oceans) by specialized bacteria and other microbes. So is the denitrification process that converts rN back into N_2, atmospheric nitrogen. The decomposition of dead organisms that releases essential nutrients for biological use as well as keeping their cycles turning is conducted by microorganisms. Consider for a moment two extremes – a healthy soil with plants, earthworms, insects, and microorganisms and another sterile soil without the organisms that typically inhabit soils. The first soil we know provides numerous ecosystem services. What about the second?

Let us begin with fixation of nitrogen by legumes (soybeans, e.g.). In a healthy soil, leguminous plants are able to fix nitrogen because **Diazotrophs**, which includes microbes such as **Rhizobia, Frankia**, and **Azospirillum bacteria**, that reside in soil take residence in root nodules of these legumes (Franche *et al.*, 2009; Santi *et al.*, 2013). These bacteria have a symbiotic relationship with the legumes where they process N_2 into ammonia that can be used by the plant for its growth and development, and the plant supplies them with carbohydrates. Sterile soils will not contain these bacteria, and unless steps are taken to inoculate the soil with the necessary bacteria, the legumes will be unable to benefit from **nitrogen fixation**.

Next in our examination is decomposition and the processing of the carbon, nitrogen, phosphorus and other elements in dead organisms. Again, we will begin with the healthy soil. Water plays a role by removing soluble compounds, leaching them into the soil. Insects, earthworms, snails and other animals aid the process by breaking larger pieces into smaller pieces. **Fungi**, and microbes will break down the remains of deceased plants and animals into their chemical parts. This breakdown involves multiple chemical processes essential for the carbon, nitrogen, and phosphorus cycles discussed earlier. As organic matter decomposes, carbon is emitted as carbon dioxide, as well as being integrated into soil organic matter. Nitrogen and phosphorus in the decomposing organic matter are immobilized as they are converted to organic forms in living organisms and **mineralize**d as these organisms die. They thus become available in the soil. Nitrogen and phosphorus are mineralized into many forms depending upon the soil moisture, temperature, oxygen available, acidity, and other variables. The immobilization – mineralization process acts to store these nutrients in the soil matrix and maintain a consistent carbon–nitrogen–phosphorus ratio. In sterile soil, on the other hand, water leaches soluble compounds as the physical degradation processes cause disintegration of leaves and other vegetative materials. In the sterile soil, this disintegration will be slower than in the healthy soil because insect, earthworm, and other animal populations will be lower. The chemical decomposition of these materials will take place at a lower rate because the microbial organisms and fungi are not present to break down the organic material into its chemical components.

Finally, we need to repeat ourselves and mention the denitrification process – the conversion of rN into N_2, where the nitrogen cycle returns to where we started. Denitrification occurs through the activity of denitrifying bacteria. Without them, our cycle is not complete and issues associated with increasing amounts of rN are exacerbated.

Admittedly, the contrast between a sterile soil and a healthy soil is extreme. We should recognize that organisms if given an opportunity will migrate into the sterile soil and begin to serve their functions. Over time plants will grow, plant roots will die, microbes will do their thing, and the sterile site will become populated and contribute to keeping the cycle intact.

4.8 SUMMARY

The reader should note that in the timeframe of a human life or multiple generations of humans, soils are non-renewable. They require thousands of years to produce a tenth of a meter (Birkeland *et al.*, 2003; Eni School, 2020). Their non-renewability argues for conserving the quality of these soils. Maintaining their biological productivity is an essential goal for anyone concerned about the environment.

The texture and structure of soils determine the flow of the ecosystem services that the land can provide. But soils do more than just predicate the behaviors of the plants, animals, and water resources at their surface. They serve as the medium in which our agricultural plants grow and in which their roots receive water and nutrients. Soils are nature's recyclers, conserving the scarce building blocks of life. They hold water and regulate its release. They store carbon that otherwise would exist as carbon dioxide, in our atmosphere. Soils thereby help regulate carbon availability, consequently essential to the functioning of systems in the biosphere in which we humans inhabit, live, and depend for our existence. They maintain the balance between the nitrogen, phosphorus, and carbon, the components in our biosphere.

Why is understanding the relationship between the ratios of carbon, nitrogen, and phosphorus to the chemical cycling and balance important for you in understanding environmental and resource management? It is because this knowledge gives you a tool for evaluating the health of ecosystems and predicting environmental outcomes. If you are told that soils are carbon poor or there is a high concentration of nitrogen in the soils, there is a good bet that the soils are in poor condition, are degrading, and will continue to degrade until the proper ratios of carbon to nitrogen to phosphorus are reestablished. The nitrogen and phosphorus will be released into water and air. Conversely a remediation strategy could be to increase carbon available to the microorganisms of soils, such as through plant residue.

Finally, scientists are still discovering all the subtle ecosystem services and benefits that soils provide humans. In today's industrial and postindustrial society, soils do more than just provide a source of bioavailable nitrogen and phosphorus, they also serve as the vehicle for eliminating the waste that we generate in our daily lives. Sludge from water purification plants is disposed of on soils, eventually becoming integrated into soils. Livestock waste is disposed of on soils, also over time becoming part of soils and contributing to the carbon, nitrogen, and phosphorus stored there. In turn, the bacteria and other microorganisms in animal and human waste re-inoculate the soil keeping its biota diverse, alive and thriving. Imagine what our cities and communities would look and smell like if soils were not available for the elimination of our organic waste.

Part II

Stuff Happens and for Every Action There Is a Reaction

Chapter 5

Natural and human-induced change

Change is the only constant in life.

> — Heraclitus – Greek philosopher

To every time there is a season, and a time to every purpose under the heaven:..

> — From Turn! Turn! Turn! A song by Pete Seeger with lyrics taken from
> Book of Ecclesiastes, King James version of the Bible

Up to now we have examined water, carbon, and nutrients as they move through the different stages of their cycles. We now consider what happens with disruptions to these systems and thus to the equilibria among carbon, nitrogen, and phosphorus pools. We must, of course, take into account that these systems are dynamic, meaning that there is always some change. Over time natural forces have moved continents, formed and eroded mountain chains, and the world climate has shifted between ice ages and tropical periods. For this discussion, the changes that we will examine are those that occur over shorter expanses of time due to natural phenomena or from human actions or influence. Whether the disturbance or change occurs as a consequence of natural flux or human action is irrelevant. The resulting adjustment is the same. We are going to look at these together and think through what happens when these changes take place.

A major driver of change is growth in the human population, especially in the past one hundred years, as shown in Figure 5.1. In 1800 the human population was approximately one billion, it was 1.6 billion in 1900, and (United Nations,

© IWA Publishing 2021. A Guide to Understanding the Fundamental Principles of Environmental Management. It Ain't Magic: Everything Goes Somewhere
Authors: Andy Manale and Skip Hyberg
doi: 10.2166/9781789060997_0077

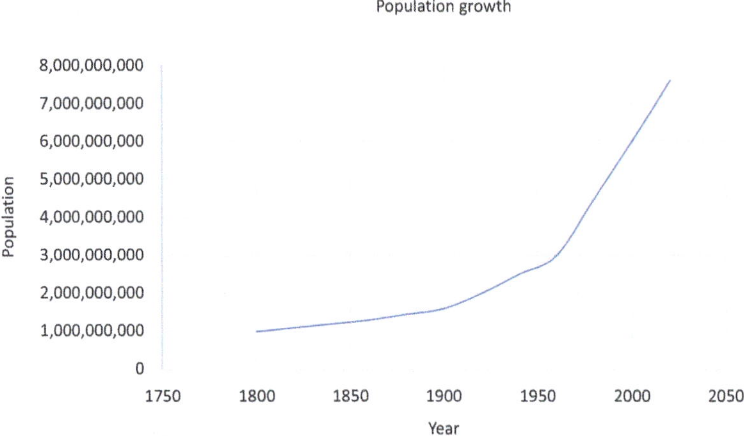

Figure 5.1 Human population 1800 to present (*Source*: Worldometer, 2020).

1999) 7.67 billion in 18 August 2020 (United States Census, 2020). More people need more food. This in turn leads to more land needed for agriculture unless there is either a profound change in diets or an increase in agricultural productivity. Increased population also means increased demand for water, shelter, and transportation. These in turn lead to more buildings and infrastructure as well as the movement of materials from one location to another. Although we will not be discussing human reproduction or population dynamics, one must recognize the importance of the role of increasing human population in forcing activities to meet these demands. In the discussion that follows we provide you a glimpse into how human activities affect the natural cycles.

5.1 NATURAL VARIATION

Change happens. The world is not at rest. Our immediate environment varies constantly. We all observe that weather differs from day to day, season to season, and year to year. Precipitation events vary from fog, to drizzle, deluges, hurricanes, cyclones, and monsoons, and reduce the temperature to below the freezing point for water and we get snow from flurries through blizzards. On the other side of the coin, we observe a gradation of dry periods up to extreme drought. We view these fluctuations in temperature and precipitation as natural.

Because we all have all observed changes in the weather, weather is a good place to start in our discussion of natural variation and how to assess it. Typically, in the absence of additional information on what future weather conditions will be, the long-term mean temperature and precipitation for a given period would be reasonable estimates for the temperature and precipitation we would expect to occur. However, because we know that weather is variable, we are not shocked if the expected value is incorrect. Still, we expect the actual values to fall somewhat close to the long-term average. Because weather varies from day to day and even more so over longer periods of time, we observe events that are decidedly unusual, such as **heat** waves, downpours, or droughts. Even though they are unusual, we also take these events in stride. Then there are extreme events – weather phenomena that fall outside our realm of experience when the precipitation and temperature are far greater or less than average. Such events include hurricanes, cyclones, extended droughts, and deluges (see Measuring extreme events box).

The natural variation in the frequency and intensity of precipitation events has a big impact on water movement. The more frequently intense storms occur, the more water that must move through the hydrological cycle as discussed in Chapter 2. At greater and greater volumes, some pathways within the cycle, such as evapotranspiration and infiltration, become saturated and the water flows into secondary routes in the **hydrologic cycle** (Kim *et al.*, 2019; Sullivan, 2020). The result is that, at a given amount of precipitation, more water will flow into a waterway than can be contained within its banks. Not surprisingly, flooding occurs. Floods and hence flooding in a landscape relatively undisturbed by humans are natural events that occur with discernible frequency, though not necessarily on a predictable schedule. Over long periods of time (measured in geologic eras), these flood events carve out plains along a river or stream. These areas along rivers and streams are floodplains, and as the word suggests they are the lands that contain the waters that overflow the banks of streams and rivers during extreme events. Although floods are natural events, human activity can change the frequency of floods and volumes discharged during floods. Damages to human health and welfare that result under these altered riverine conditions are therefore not entirely natural phenomena outside of human control. Human activity can influence the damage resulting from flooding. In this chapter, we will use our understanding of the hydrologic cycle, our confidence that everything goes somewhere, and the uncompromising effect of gravity to examine how human activity can affect the frequency, severity, and damage of floods.

Droughts, which are on the flip side of precipitation events, are also natural events. They occur during periods of reduced precipitation. Natural variation in precipitation over longer periods of time implies that there will be both surpluses and deficits in rain and thus floods and droughts. Droughts affect plant growth and the availability of water for all fauna, humans and wildlife alike. Interestingly, droughts, while not as photogenic as floods, generally affect larger areas and cause more economic disruption (Lippsett, 2012; NOAA, 2020). As we will see, human activity can increase the severity of droughts. We will discuss in detail these relationships.

A third natural phenomenon is fire. Fire occurs naturally when fuel, **desiccation**, and ignition occur. Because each of these elements are not present in all years, fire occurs sporadically. Fuel is generated by vegetative growth and can accumulate over time. Fire consumes the fuel available thereby reducing the likelihood of another fire until the fuel is replaced. Obviously, fire is more likely when the fuel is dry, a condition that occurs in periods with low precipitation, and becomes particularly acute during droughts.

Fire plays a critical role in shaping and maintaining ecosystems (Bailey, 1988; Williams, 1995). Fire releases nutrients and carbon into the environment and has the ability to change vegetative covers. Frequent less intense fires can help maintain the health and vigor of ecosystems, while occasional intense fires can expose landscapes to erosion and degradation, permanently damage soils, and alter vegetative covers (Bailey, 1988; Williams, 1995).

5.2 ANTHROPOGENIC DISTURBANCE: CONSIDERING HUMAN EFFECTS

Why do we differentiate **anthropogenic** change from natural variation? Is it because change is bad? Obviously not because change occurs in natural systems. It is because human-induced change can alter the mass and relative proportions of the water, nitrogen, phosphorus, and carbon in the pools we have examined. This disrupts the balance between and among pools setting in motion a rebalancing. Systems react. Our goal here is twofold: to help you distinguish between natural variation and change from human actions and to show how these actions can cause disruptions. By understanding how cycles move to reestablish balance, you can anticipate how ecosystems can react. This is at heart of environmental management.

Measuring extreme events

Extreme events are identified using statistical analysis of past occurrences. In general, the mean is identified, and the variance around the mean is used to identify the likelihood an event will occur in any given year. Some events will be common and expected to occur each year, while others will be unusual or even rare.

A storm that has a 1% chance of occurring in a given year is termed a hundred-year storm and one that has a one-tenth of 1% chance of occurring in a year is a thousand-year event.

Hydrologists use their knowledge of how a watershed will react to a storm to estimate the peak flow from that storm. It is dependent on both the precipitation from a particular storm and precipitation from earlier storms. The precipitation that precedes a new storm is a factor because it affects the stream level as well as how much water can infiltrate.

Every year holds the same chance of the event. An occurrence in a prior year has no effect on the likelihood of occurrence in the next or following years.

It is important to recognize that there are many watersheds across a country such as the United States. With a large number of possible events, numerous hundred-year floods are likely to be observed each year.

How do people alter the balance? What are the human activities that affect the quality and integrity of our environment? People have obviously modified landscapes, converting grasslands, forests, and wetlands, into cropland, pasturelands, cities, and highways. We have changed the path of waterways, dammed rivers and created reservoirs, harvested forests, drained wetlands, and irrigated deserts. In so doing, we have altered the hydrology of watersheds, water quality and availability, the balance of carbon, nitrogen, and phosphorus in terrestrial pools, and introduced material (mined or excavated) that were in long-term earth storage. The mining and drilling of fossil fuels (coal, natural gas, and petroleum) and their subsequent combustion released vast amounts of carbon from the crustal pool into the atmospheric pool as carbon dioxide. Using long-buried deposits of guano as a fertilizer freed reactive nitrogen and phosphorus to enter air, water, and soil resources. Phosphorus as phosphate has been mined from phosphate rock, particularly apatite. And more recently, industrial processes have made reactive nitrogen plentiful.

The altering of the natural cycles of carbon, nitrogen, and phosphorus and the harnessing of water for agriculture have been associated with a five-fold increase in human population since the invention of modern agriculture and a near doubling of the human population in the past fifty years.

5.3 THE HUMAN FACTOR

Modern humans have changed the relative dominance of different pathways within carbon and nutrient cycles, making what was once scarce plentiful, shifting carbon, nitrogen, and phosphorus from long-term storage pools into short-term pools and thus bioavailable. We used these newly available compounds to produce the crops we depend upon for our food, fiber, and fuel. Initially, this was done by our ancient ancestors recognizing the relationship between certain practices and greater production (Balter, 2015). Our early farming forebears thousands of years ago began to till the soil, thereby fertilizing the crops by releasing the reactive nitrogen and phosphorus for uptake (while turning the carbon it into CO_2) and loosening the soil for root growth. They did so because they saw the plants grow faster and bigger. Even the byproducts of human and livestock production, human and animal waste was recycled, whereupon they were deposited on crops. Again, these early farmers saw the crops respond by growing greener and bigger. The nitrogen and phosphorus that was made available entered the flux of material cycling among their storage pools. In these ways, humans altered the nitrogen, phosphorus, and carbon cycles and began human efforts to manage the environment.

There are a number of ways in which humans have altered long- and longer-term storage pools of carbon, nitrogen, and phosphorus. Remember that the ratios of carbon to nitrogen to phosphorus in soils are roughly constant. [They are also in constant ratios to each other in ocean systems, though the mechanisms by which these equilibria occur are different.] A change in the amount or proportion of the one results in changes to the others. Because it is so basic to human existence and most people are broadly familiar with agriculture, we are using agriculture to demonstrate how these relationships function. Plowing or tilling disturbs the soil, oxidizes organic material, making nitrogen and phosphorus available for crops and releasing carbon into the atmosphere as CO_2. The CO_2, nitrogen, and phosphorus in turn become available for reuse or, in the case of excessive disruption, transported away through water movement.

The draining of wetlands and floodplains benefits people by enabling these lands to be used for the purpose of agriculture. The water-saturated nature of these soils permits carbon and hence nitrogen and phosphorus stocks to build up over large expanses of time, sometimes resulting in soils dozens if not hundreds of feet deep. Draining them exposes the soils to oxygen and therefore **oxidation**. This releases CO_2, rN, and bioavailable phosphorus. These organic nutrients are available for crop production. What remains can escape into the environment.

Farmers also discovered that rotations can improve productivity. They found that including legumes in **rotation** (the growing of crops in the subsequent growing seasons or seasons) with other crops or including a fallow period improved productivity. We now know that legumes have symbiotic relationships with microorganisms that can obtain reactive nitrogen from the atmosphere. This rN from the nitrogen-fixing crop also fertilizes the subsequent nitrogen-demanding crop grown in the same soils. Fallow periods permit soils to replenish nutrients and store moisture, as well as breaking up disease and pest cycles and allowing soil biota to reestablish (Padgitt, 1997).

Early humans discovered that recycling their own and their animals' waste products and incorporating them into agricultural lands increased yields by renewing the soils. In much of the world, human waste has been collected as 'night soil' and then returned to the land as fertilizer or compost. Livestock waste, even in industrialized facilities, is still collected and returned to the land as fertilizer. Still, because the delivery of these recycled nutrients is not 100% efficient, some portion of the nitrogen and phosphorus is lost to the atmosphere and water. This loss can be because the nitrogen or phosphorus is in a form that cannot be utilized, such as nitrous gas, or because the dissolved nutrient bypasses the plant (e.g. phosphate and nitrates draining into groundwater). Coupled with the loss of carbon from tillage or excessive drainage, agricultural soils have tended over time to become stripped of their nutrients and less productive. These conditions require new sources of nitrogen and phosphorus to be added in order to maintain soil productivity.

In the nineteenth century, the most important source of fertilizer was guano, as mentioned previously, the excrement of millions of birds that has accumulated over hundreds if not thousands of years. High in nitrogen, phosphorus, and other key nutrients, guano was highly sought after as a fertilizer for crops. Guano, however, is not found everywhere. The main sources are located on islands in South America. The demand for this rich source of fertilizer helped drive European and American colonialism because of the importance of fertilizer in the production of food for an ever-expanding population (Heimlich & Daugherty, 1991; Davies, 2019; Immerwahr, 2020).

5.4 ARTIFICIAL FERTILIZER

In this the **Anthropocene Epoch**, humans have not just transformed their world in an obvious physical manner, but also in chemical and biological ways (Crutzen, 2002; International Geosphere-Biosphere Program, 2020; Pennisi, 2020). Our cities and roads are the most obvious examples of the former; our atmosphere, lakes, and rivers, the latter. Both transformations have profound effects, but it is the latter that can have the most enduring impact on the world that we know, with grave consequences for future generations.

Anthropogenic activities have altered the chemical balance for elements that become the building blocks of our living environment. The increase in carbon dioxide levels in the atmosphere with its concomitant manifold impacts on climate and weather, sea levels, biologic diversity, and ocean acidification is an example that is familiar to all of us. Less popularly known and appreciated is how we humans are altering the nitrogen and phosphorus balances relative to the concentrations of their myriad reactive forms and to what is not biologically available.

The change in the percent of nitrogen converted from inert to reactive is small, but because reactive nitrogen can be stored biologically and assume many forms in nature before it is returned to rest again as N_2, the terrestrial total can accumulate over time. Scarcely more than a hundred years ago, a German chemist by the name of Fritz Haber discovered a process by which to take the nitrogen in the air and convert it to reactive nitrogen. In 1910, Carl Bosch developed a method that permitted this process to be undertaken at an industrial scale. In so doing, they largely eliminated the nitrogen constraint to increased agricultural production. Until then, farmers were constrained by available nitrogen and, to a lesser extent, phosphorus. Soil stores of nitrogen and phosphorus were used, reducing the availability of these essential nutrients for future vegetation. And wars were fought in part over food insecurity resulting from a constraint on production posed by the scarcity of nitrogen. Humans, since the Haber–Bosch process was discovered, have greatly altered nitrogen balances. Now far more reactive nitrogen is manufactured (and generated by fossil fuel burning) (see Figure 5.2) than produced by natural processes. Nearly half of the world's population now depends upon artificially created reactive nitrogen for its existence (Smil, 2001).

Before the early twentieth century, most of the available reactive nitrogen for use in agriculture came from soils, animal and human waste, or lightning. Since the Haber–Bosch process was discovered, humans have doubled the amount of new reactive nitrogen introduced into the biosphere each year, from roughly 203 Tg (teragrams or a million times a million grams, or 1 TG = 1,102,311 U.S. tons) per year in preindustrial times to what is roughly 413 Tg per year (Fowler *et al.*, 2013). About one-seventh of the increase is the result of the combustion of fossil fuels for energy and transportation (Fowler *et al.*, 2013). More than 120 Tg

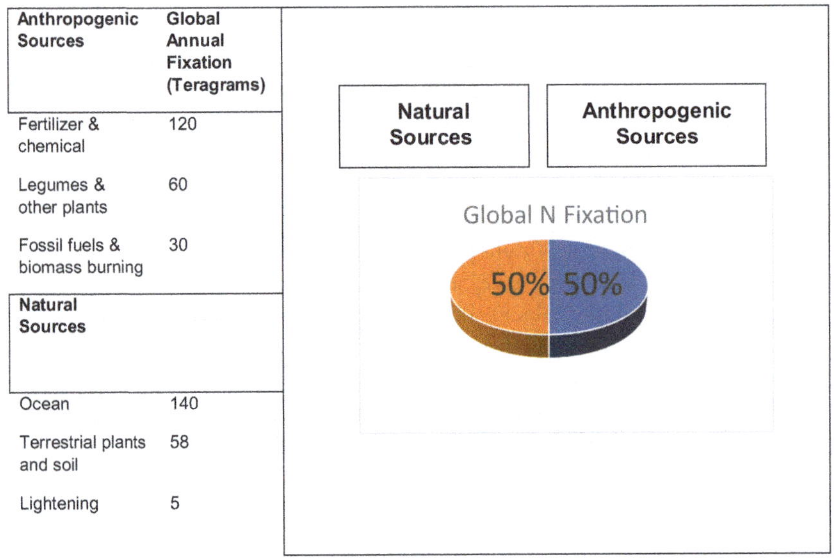

Anthropogenic Sources	Global Annual Fixation (Teragrams)
Fertilizer & chemical	120
Legumes & other plants	60
Fossil fuels & biomass burning	30
Natural Sources	
Ocean	140
Terrestrial plants and soil	58
Lightening	5

Figure 5.2 Sources of reactive nitrogen (*Source*: Fowler *et al.*, 2013).

comes from fertilizer (Fowler *et al.*, 2013). An additional 60 Tg comes from the introduction of legumes into agricultural rotations to fix nitrogen for crops (Fowler *et al.*, 2013). Figure 5.2 shows how humans have drastically altered the nitrogen cycle. Of the amount of additional, synthetic nitrogen for agriculture, a small portion or roughly some 10–15 Tg is consumed as food (based upon Smil, 2012). The rest is displaced into the environment, into water in the form of excess nitrates, and the air as nitrous oxide and ammonia.

Global production of new reactive nitrogen for industrial and agricultural purposes increases nearly 1% per year (Food and Agriculture Organization, 2017). Remember that whatever new reactive nitrogen that was introduced into the environment in a previous year <u>and</u> was not converted to the long-term atmospheric, ocean sediment, or soil pools remains in the biosphere, compounding whatever environmental problems with rN may exist. Figure 5.3 shows how the amount of rN introduced by humans has increased rapidly over the past century and is expected to continue this increase for decades to come.

People's appetite for rN is expanding as world population expands and societies, in general, get richer. Not surprisingly, with rising incomes comes greater demand for meat. In the United States, some 2.6 kilograms of grain is consumed by beef cattle, to produce a one kilogram of beef, as presented to the consumer in the supermarket (Suszkiw, 2019). Farmers use 1 kg of nitrogen to produce 65 kg of grain, which means 1 kg of beef requires 0.04 kg of nitrogen ($2.6 \times 1/65 = 0.04$ kg) (U.S. Environmental Protection Agency, 2011). If we eat a hundred

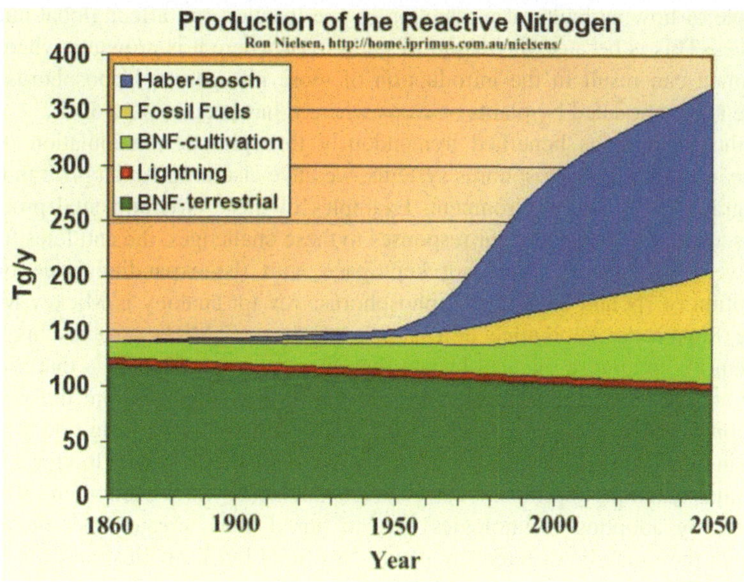

Figure 5.3 How humans have affected the N cycle (*Source*: Nielsen, 2005).

kilograms of beef per year, we are consuming some 4 kg (100 × 0.04 kg = 4 kg) of rN that had to be used to produce our dinners. As more people on the planet achieve middle-class status, more meat is consumed. The result is the rapid rise in the production of rN.

New phosphorus, the other key fertilizer, is mostly mined. This mining of phosphorus removes it from the longer-term storage pool, as described in Chapter 3, and moves it to the bioavailable, surface terrestrial pool. In 2019, some 46 thousand metric tons (50.7 thousand U. S. tons) were produced. In 2015, the figure was less than 42 thousand metric tons (46.3 thousand U.S. tons), an increase of almost 10% in just four years (Food and Agriculture Organization, 2017).

5.5 TRANSPORT OF NITROGEN AND PHOSPHORUS THROUGH GRAIN, LIVESTOCK, HUMANS, AND SLUDGE

For a problem to be global, it must either have local impacts that are suffered globally or globally distributed impacts. **Air emissions**, such as nitrous oxide that distribute themselves globally are clearly a global issue. But what about a local activity that primarily affects land and water? Even site-specific application of fertilizers to crops precisely measured to meet crop needs can still result in an environmental imbalance because some of these nutrients leave the site. But beyond this the production, harvest, and transport of grain provide a good example of how agricultural production in one location can affect global nutrient balances. This is because the export of grain from where it is grown to where it is consumed can result in the introduction of more nitrogen and phosphorus than can be accommodated by plants or crops where it has been imported.

Although humans benefited tremendously through our manipulation of the carbon, nitrogen, and phosphorus systems, we have also long paid a price in terms of degradation of our environment. Examples of these environmental problems are discussed in Chapter 6. Our responses to these challenges, the antidotes for the negative consequences, have not kept pace with the expanding scale of the generation of rN and bioavailable phosphorus. An apt analogy is Mickey Mouse as the Sorcerer's Apprentice in Disney's "Fantasia." Mickey solves his work problem by creating a magical broom that does the work for him – that is carry water and fill the basin. Unfortunately, he fell asleep and therefore did not pay attention to when enough is enough. His attempts to quell the rising tide of water result in more brooms, only making the problem worse, with more flooding.

Human reactions to problems can be much like Mickey's. We too often solve one problem by adopting technologies that, to our dismay, create other problems, although they may be different by nature. Artificial fertilizer illustrates this serial problem. Fertilizer use has improved the health and wellbeing of billions of people on this planet, yet the overflow, the increased availability of rN now poses a threat to many important ecosystems essential to human existence. The existential problem with nitrogen to be addressed is the imbalance between the global rate of production of reactive nitrogen and its rate of immobilization or removal and conversion back to N_2.

To examine how local agriculture can lead to global imbalances, let us first ask a couple of very basic questions: Are there a lot of agricultural products imported or exported? If so, does what is transported represent much nitrogen and phosphorus?

We will focus on grain because grain export constitutes the most trade in raw agricultural products. More specifically, let us use corn as an example, because corn is widely traded worldwide and corn production is a major user of fertilizer. Similar analysis could be conducted examining other grains and meats. According to U.S. Department of Agriculture's Foreign Agricultural Service (2020), roughly 171 million metric tons (188 million U.S. tons) of corn were traded in 2019. The corn was harvested from a little more than 190 million **hectares** (409 million acres) of land. The United States accounted for 33 million hectares (81.5 million acres) of corn harvested and roughly 49 million metric tons (54 million U.S. tons) exported. Because only the kernels are traded, we examine only the nutrient content of the kernel, not the corn plant. The mean nitrogen content of the corn grain is roughly 1.54% (Boone *et al.*, 1984). For phosphorus, the values can vary considerably, but for the sake of this analysis, let us use the average phosphate content of 0.195 kg of phosphate per bushel (or 0.43 lb.). There are 39.368 bushels per metric ton (MT) of corn (U.S. Grains Council, 2020). Doing the math

(0.195 × 39.368) shows that there are 7.7 kg of phosphate per MT of grain (Nafziger, 2017). Multiplying the numbers gives us an estimate for global transport of 1.31 million metric tons of phosphate and 2.63 million metric tons of nitrogen. (Global phosphate transported in corn = 171 million MT × (7.7 kg/MT)/(1000 kg/MT) = 1.31 million MT; Global nitrogen transported in corn = 171 million MT × 1.54% = 2.63 million MT.)

This is a large amount of nitrogen and phosphorus, but do we know whether or not it is large enough to affect anything? Let us compare it to what occurs naturally. Lightning fixes between 3 and 10 Tg per year. Since a teragram is equivalent to 1 million metric tons and using the midpoint, lightning fixes 6.5 million MT (7.1 million U.S. tons) of nitrogen. Therefore, we see that trade just in corn moves 2.63/6.5 or about 40% of what lightning provides in new reactive nitrogen globally. For phosphorus (i.e., phosphate), let us compare our estimate to the global consumption of 47 million MT (52 million U.S. tons) in 2018 (USGS, 2019). Thus, about 1.31 divided by 47 or roughly 2.8% of annual global production of phosphorus as phosphate is exported as grain (USGS, 2019). Are these numbers significant? In considering your answer, remember that the transport of these quantities of nitrogen and phosphorus is (1) going to concentrate the nitrogen and phosphorus in areas with either concentrated human or animal populations, (2) in locations that are likely to lack the capacity to process the additional nitrogen and phosphorus and (3) this transport occurs each year.

Most of the nitrogen and phosphorus will initially remain in the destination where it is consumed and transformed into animal or human waste. As explained earlier, this animal waste (along with human waste that has been processed through sewage treatment plants) will ultimately be disposed of, that is added as a fertilizer, on cropland U.S. EPA, 2020a). Because the ratio of nitrogen to phosphorus in human and animal waste is not consistent with their corresponding ratios in soil, their application without further processing to resolve the imbalance can result in too much of one or the other being applied. This can overload the capacity of the soil microbial systems to convert the nitrogen and phosphorus into new soil leading ultimately to the transport of nitrogen (and under certain circumstances, phosphorus) offsite into the air and/or water. The volume of waste makes the potential for overloading the soil capacity a concern in many areas. Unless, there is sufficient cropland or other vegetated land to accommodate the waste generated by large numbers of animals and humans, and unless the **manure** is applied at appropriate rates, there will be residual nutrients that can move offsite from land to water and/or air.

The amount of nutrients transported as animal products can be estimated by working backwards from the type and number of animals (generally, chickens, pigs, and beef and dairy cattle). The conversion rate, that is the ratio of grain to animal weight, for chickens is roughly 2 to 1. For pork, the number ranges from 3 to 1, to 4 to 1. And for beef the number is between 7 and 10 to one (Oros, 2020; Wikipedia, 2020). The grain that is not incorporated into the animals is expelled as manure, and the manure contains a large amount of nitrogen and phosphorus.

Traditional farming practices that cycled nutrients locally from fertilizer, to crops, to grain, to animals that generate waste that is then returned to the land as fertilizer have been disrupted. Modern agricultural systems still cycle nutrients, from fertilizer to grain, meat, and manure, but the nutrients in grain and meat are transported, making the geographic area in which cycling occurs larger. Efficient use is correspondingly more difficult. Animal and human waste generally has a large water content which makes the waste heavy and expensive to transport any significant distance. The historical rule of thumb has been the following: for manure to be applied in an economically feasible manner (it needs to be applied within 40–50 miles (64 to 80 km) from where it is generated).

Even poultry waste which has a low water content and thus is relatively light given its mass of nitrogen and phosphorus is generally not transported more than 50 miles or 90 km from where it was produced. Poultry waste can also be burned and used as a fuel source, emitting its carbon into the atmosphere, but leaving the nitrogen and phosphorus as residue which can then be stored and transported at lower cost. However, care must be used when applying the waste as a fertilizer because of the relatively high phosphorus to nitrogen ratio and the variability in this ratio between different sources of chicken litter.

Regional and global transport of grain and the cycling of nitrogen (and phosphorus as well) are illustrated in Figure 5.4. The import of **feed grains** divorces livestock production from the land where livestock waste could be recycled, resulting in a concentrated mass of nitrogen and phosphorus. The residual nitrogen and phosphorus not taken up in crops, converted back to N_2, in

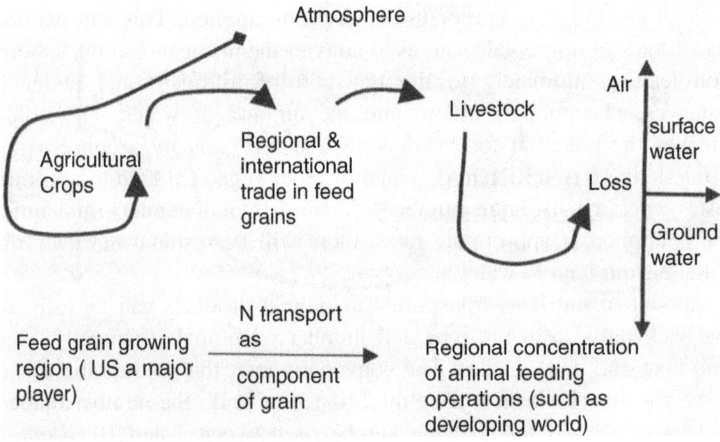

Figure 5.4 Trade impact on the N cycle.

the case of nitrogen, or incorporated into new soil is lost to the atmosphere and water.

5.6 LANDSCAPE MODIFICATIONS THAT AFFECT WATER, AND CARBON, NITROGEN, AND PHOSPHORUS

In a preindustrial world, the amount of rN and phosphorus bioavailability was constrained by natural processes. Biologically available amounts of rN and phosphorus were rapidly taken up by plants. When plants died, the now bioavailable nutrients were integrated into the soil, maintaining the ratio of carbon to nitrogen and phosphorus present. In wet soils, like wetlands or floodplains, denitrifying bacteria utilized remaining rN converting the nitrogen from its reactive form back to its stable inert form, N_2. Phosphorus, which was almost always in short supply in soil and plant ecosystems, was quickly recycled. In this way wetland, forest, and grassland soils serve as regulating media, both as sinks retaining rN and bioavailable phosphorus and, in the case of wetland soils, processing excess rN. The soils kept excess bioavailable nitrogen and phosphorus from moving into the air and water pools, and thus making them biologically unavailable.

The wetland and river floodplain soils (hydric soils) can remove vast quantities of nutrients from water. However, for these lands to be effective in denitrifying and removing excess nutrients, the soils must be in contact with the nutrient-enriched waters. Land modification such as draining and conversion of wetlands to agricultural and urban use, dredging and straightening of rivers, excessive livestock grazing, and alteration by exotic plant species reduces the efficiency of these lands in denitrification. Phosphorus from fertilizer applications or animal waste builds up in the soils unless removed by harvesting the crops grown on these soils. Oversaturation of the soils with phosphorus leads to bioavailable phosphorus becoming dissolved in soil water which can lead to the phosphorus being transported offsite.

As developed and developing nations demand more agricultural production of food, feed, fiber, and now fuel, the problem only escalates. The seemingly small changes to our ecosystems over many generations – such as the draining of wetlands, the straightening of rivers, agricultural monoculture, and confined animal feeding operations – aggregate to large imbalances of nitrogen and phosphorus. What these imbalances mean for the health of our environment and our welfare is discussed in Chapter 6.

To reiterate the message, we conveyed at the beginning of this chapter, the chemical elements, nitrogen and phosphorus, can assume many chemical forms in combination with other elements in our environment. In so doing their properties change as well as changing the media, water or air or land, in which they reside. Though we do not see or perceive these transformations, their mass, or if you prefer their number of atoms, does not change. It is conserved. In nature and in ecology, eliminating a chemical compound from one medium or one location means it exists in a different form or in a different medium or location. In managing natural resources, one must think of our chemical world in terms of availability and reactivity in juxtaposition with sinks and inertness. Everything goes somewhere.

5.6.1 Land modifications affecting the chemical balancing act

We will now identify a number of human activities that influence the movement of water and nutrients and follow the logical implications of these activities on water quantity and quality, and start considering alternatives for addressing the adverse effects of human activities. The activities we will examine are changes within the broad landscape, including vegetative cover and infrastructure (**dams**, roads, **levees**, and **irrigation** and drainage systems), and increases in nutrients and other inputs introduced into the environment.

It is important to bear in mind that most environmental management analysis involves examining changes from the current condition, not changes from natural systems. We use this reference point because many of the issues that environmental managers address are in systems that have been previously altered. Their task is often to ameliorate problems that were not anticipated when the system was modified. Examples include developing cropland conservation systems to address erosion or nutrient runoff, designing urban infrastructure to slow and reduce storm water runoff, and siting roads and other infrastructure to avoid issues with flooding, erosion, and susceptibility to extreme events.

5.6.1.1 Altered vegetation and land cover

Humans have fundamentally changed the landscape. In premodern times this would have involved changing the natural vegetation; for modern man this includes constructing buildings and other structures, paving roads, parking lots and other areas, and installing other infrastructure, but also rehabilitating degraded parks or cropland, reclaiming industrial sites, reforestation, and establishing conservation covers. What implications do these changes have for water as it moves across the land surface?

As was discussed in Chapter 2, the land cover affects the interception of precipitation, infiltration of water into soils, the speed water moves across the landscape, and the amount of evapotranspiration. In Chapter 6 we use gravity, observation, and the fact that water does not disappear to look at these individually and examine the likely impacts on rivers, lakes, and streams in the context of a series of increasingly more intense precipitation events.

Interception – Changing land cover changes interception as a rule, trees intercept more precipitation before it hits the ground than do shrubs, which in turn intercept more than grass (Dunkerley & Booth, 1999; Li *et al.*, 2017; Ufoegbune *et al.*, 2010; Sheng & Cai, 2019). Roads, buildings, and other structures are considered land surfaces (Hollis, 1988), so they do not intercept precipitation, but some water will evaporate before flowing into waterways (Hollis, 1988; Smith *et al.*, 2013). During small precipitation events, a sizable portion of the precipitation is intercepted and evaporates before reaching the land surface. As precipitation events increase in volume, the percent of the precipitation that is intercepted and evaporates decreases. Thus, the role vegetative interception plays in water movement diminishes as precipitation events increase in intensity (Dunkerley & Booth, 1999; Li *et al.*, 1999; Ufoegbune *et al.*, 2010; Sheng & Cai, 2019).

Infiltration and velocity of overland flow – Changing vegetative cover alters how rapidly water moves across the landscape as surface flow, as well as infiltration into the soil. These changes occur because modifying the vegetation both alters the roughness of the landscape and the quality and constitution of the soil (especially carbon content and bulk density) and thus the soil's capacity to hold water. Changing the roughness modifies the obstacles to water surface movement, altering the time available for infiltration, and thereby changing the amount of water that infiltrates into the soil and becomes part of the base flow. Forest and grassland soils have more pores and pore space and have a higher soil carbon content than does the same soil in cropland. Lands converted to cropland have a surface less rough than forests, grasslands, and other natural systems and generally retain less water.

Human development, such as roads, parking lots, and buildings, increases the amount of impervious surface within a watershed. If impermeable surfaces replace vegetation, infiltration is eliminated, and the amount and velocity of surface water flow increases. Water moves into streams more rapidly, contributing to higher peak stream flow that occurs with precipitation events (Hollis, 1988). Faster surface runoff has more energy to move soil and other materials.

With more impervious surfaces and a smoother landscape, single storm events can cause surface flow to reach streams more rapidly. This is because water infiltrating into the soil moves towards the stream laterally and much more slowly than surface flow. These differences are reflected in differing stream responses to storm events in altered and unaltered watersheds (revisit Figure 2.4). If the precipitation is a frozen state, it can be stored over winter. The ice either on the surface or within soils can act as an impervious layer stopping or slowing infiltration. When rain falls on melting snow and ice (the frozen remnants of previous storms), surface flow includes water from multiple events. The combination of surface flow containing water from multiple storms and reduced infiltration explains why many floods occur in the spring when spring rains falls on snow and ice from a heavy winter. Changes to the landscape can either exacerbate or mitigate this.

Changing vegetative cover affects evapotranspiration. Different plants transpire different amounts of water. When you think about it, at the extremes you already know this (consider the difference in moisture demand for a cactus and a tree). The same can hold true for grass versus trees and different crops (Sinclair *et al.*,

2005; Asbjornsen *et al.*, 2007). The rate of evapotranspiration can allow for more space for water to be retained (soil water storage capacity), and affects both the surface and subsurface flow into rivers and streams (Hatfield *et al.*, 2009).

As you consider how a changing land cover alters the quantity and velocity of water movement, remember that water is a primary agent for moving soil, nutrients, and other materials across the landscape. We will be examining some aspects of these changes in Chapter 6.

5.6.1.2 *Irrigation and drainage: two sides of the same coin*

Since the advent of agriculture farmers observed that too little or too much water harmed crop growth. If too little water was inhibiting crop growth, the obvious solution was to irrigate their cropland. Farmers began irrigating early on. 'The earliest archeological evidence of irrigation in farming dates to about 6000 B.C. in the Middle East's Jordan Valley' and there is evidence irrigation was being used around the same time in Egypt (Hoffman *et al.*, 1990; Sojka *et al.*, 2002). Today irrigation is a major tool in providing food for a world with more than 7 billion persons. 'Irrigated agriculture now covers 275 million hectares (680 million acres) – about one fifth of cultivated land accounting for 70% of global water withdraws' (UNESCO, 2019a, b). In short, irrigation is an important human intervention on the landscape.

Irrigation provides substantial benefits by increasing and stabilizing agricultural production for a rapidly expanding world population; however, by removing water from waterways, ground water, and aquifers and diverting it to cropland, irrigation modifies water regimes and can result in **soil salinization** (increasing concentrations of salts in the soil from either the irrigation water or leachate from minerals in the soil or soil column), waterlogging, and reduced water availability in the areas from where the irrigation water has been drawn. Additionally, not all of the irrigation water is used by crops. Irrigation water has greater exposure to the atmosphere leading to higher evaporation before it reaches plants (Martínez-Granados *et al.*, 2011). Residual water (irrigation return flow or tail water) that leaves the fields transports sediment, fertilizers, mineral salts, and other agricultural chemicals, and can degrade water quality.

Where there is too much water for the growing of crops, drainage systems are often used to remove excess water. These are often sites that tend to flood or become saturated during the growing season. These drains function by interacting with the water table to provide an unimpeded path for water to flow into waterways (Royer *et al.*, 2006; Ahiablame *et al.*, 2011; Williams *et al.*, 2015). They lower the water table providing aerated soils more conducive for root growth. Drainage systems can be shallow ditches that carry water off a field, extensive, highly engineered subsurface drains removing water from substantial **catchments**, or the windmills and powerful pumps that have reclaimed the Dutch lowlands.

Drainage systems alter the hydrologic nature of rivers and streams by moving water more rapidly into streams, intercepting water tables and water infiltration, reducing subsurface base water flow, and lowering the water table. The more rapid movement of water into streams during and immediately after a storm event results in a higher stream **peak flow**, while the reduced subsurface recharge causes lower stream base flow between storm events (Schilling & Helmers, 2008; Smith *et al.*, 2013). By changing how water moves across and through the landscape, drainage also affects the transport of soil and nutrients.

There are, in general, two types of drainage: **surface ditches**, which intercept ground water tables and drain to waterways, and subsurface drainage. Subsurface or tile drains are perforated pipes generally buried 3–6 feet underground that drain into waterways (Panuska, 2017). The 2017 Census of Agriculture (National Agricultural Statistics Service, 2019b) reports that there are 56 million acres (23 million hectares) of cropland with **tile drainage** in the United States and 44 million acres (18 million hectares) of surface drainage cropland. Tile drainage systems are concentrated in the Midwest where corn and soybeans are the predominate crops.

5.6.1.3 Levees

Levees are embankments along the sides of a river built and placed to protect buildings and cropland by keeping flood water from leaving the channel and flooding the site. Levees alter river flow and in general disconnect the river from natural floodplains, which would store water during flood events. Too often levees succeed by causing water to flow into areas where it normally would not. Thus, if land on both sides of a river are of equal height but one side has added a levee, the water on the other side will flood higher than if the levee were not there.

If both sides of a river have levees, then the river will be narrower at that location than it would have been during flood events; and water will be forced downstream. Because the same amount of water is trying to move downstream through a narrower passage, the water level will rise and the rate of flow will be faster. This results in higher peak flows during floods for downstream communities. The greater rate of flow concentrates the energy of the river increasing stream bank erosion and the scouring of the river channel. Faster moving water also means that the water has more energy to move things such as sediment, changing the river morphology as less sediment is deposited upstream and more is carried downstream. The erosion and scouring result in sediment and the materials attached to it being brought into the water column, and transported downstream. The sediment and the nutrients attached to the sediment are deposited downstream as the current slows.

In many high-water events, levees are effective in moving water past protected sites to areas designated to receive flood waters. However, under extreme flooding events, levees can be overwhelmed, resulting in catastrophic damage to affected communities. Another effect of constraining rivers is that an elevated water level on the main stem of a river can, during flood events, prevent water from draining from the tributaries into the main stem. Flooding in these backwaters can be significant (Singh, 1996). Straightening rivers (eliminating the natural curving of rivers in relatively flat landscapes) increases flow and the amount of sediment scoured from river beds and banks.

5.6.1.4 Dams and reservoirs

Dams are structures placed along waterways to raise the water level and create a **reservoir**, as well as to regulate flow. Reservoirs store water from river flow. They temporarily hold water from stream flow and release the water in a managed manner. The water leaving the dam is called outflow. Dams can provide some control over river flow and have been used to reduce flooding and provide minimum flows during droughts. However, the variable river flows from natural climatic variation means that dam managers need to determine how much outflow to release and how to schedule to releases. These decisions determine river flow below the dam and require difficult decisions that involve conflicting objectives, placing a great deal of responsibility on those managing water. These tradeoffs can become acute during extreme events such as flooding and drought. Examples of conflicting objectives include water storage, irrigation demand, flood control, river navigation concerns, energy generation, and the health of riverine ecosystems.

Dams block stream flow causing suspended sediment to settle so that over time, these reservoirs collect silt and sediment that otherwise would flow downstream, thereby decreasing their potential for mitigating peak flows during storm events. Larger sediment particles moving downstream normally replenish lands, such as wetlands, located in or adjacent to the river. In areas where the land is sinking due to geologic phenomena, this loss of replenishment by heavier sediment leads to their disappearance (Syvitski & Kettner, 2011). Many, if not most, dams constructed in the United States for the purpose of mitigating downstream flooding have reached or are approaching their maximum operational lifespan given when they were constructed – most over fifty years ago. Dam failures or sediment-filled dam reservoirs can cause catastrophic damage downstream.

Reservoirs can provide multiple services such as energy production, urban water supply, irrigation, and recreation. The classic example for energy generation is the old mill pond used to drive the mill for grinding grain, while the more modern example is the hydroelectric plants in western North America. Reservoirs provide greater stability for human water supplies, but increase water loss to evaporation, and disrupt river hydrology. These disruptions can cause environmental problems that may not manifest themselves for decades or even longer by changing the amount and type of sediment that rivers otherwise carry downstream, altering water temperature that can affect aquatic life, and affecting the ebb and flow of water levels and hence fish and animal habitat. Finally, all reservoirs behind dams eventually fill up with sediment, gradually losing their capacity to store water and creating a potential threat to the structural integrity of the dam.

5.6.1.5 Livestock production and manure

Humans have domesticated numerous animals. Cattle, hogs, horses, chickens, oxen, sheep, and many other animals have for millennia been breed and raised for food and labor. Among the consequences of the nexus between humans and livestock are improved diets, the need to feed these animals, and the production of animal manure. Although initially farmers did not have a modern understanding of the **chemical** processes involved, they observed that adding manure to soils increased crop production. We now know that the nitrogen, phosphorus, and carbon contained in manure provide nutrients and enhance soil health. Additionally, manure stimulates microbial populations that facilitate soil functions and enhance soil health.

Although manure contains nitrogen and phosphorus in plant available forms and has value as a fertilizer, the proportion of these elements varies by livestock type and fluctuates as much as plus or minus 30% due to genetics, diet, mineral supplements, farm management, and other factors (Lorimar *et al.*, 2004). The great variability in manure nutrient composition, diverse soil characteristics, and different crop nutrient requirements mean that the nutrient content rarely contains an optimal mix of nitrogen and phosphorus. This variability poses a dilemma for farmers because it makes it difficult to apply manure at a rate that meets plant needs and avoids applying excess amounts of nutrients that can pollute waterways. For this reason, the nutrient content of manure should be tested (Dou *et al.*, 2001).

Early farms were small and often raised both crops and animals, a production system that facilitated the use of manure with few environmental issues. However, as modern agriculture developed, farms became larger and more specialized. In more developed countries, livestock production enterprises (used in this text to refer to all animal agriculture) have tended to expand in size and become highly specialized. Frequently they do not produce crops that could receive manure as fertilizer (Kellogg *et al.*, 2000; Ribaudo *et al.*, 2003a, b). Manure is applied (and often overapplied) on adjacent land as a waste by-product [see What is waste? Box] rather than a resource for crop production.

The chemical and hydrological processes discussed in earlier chapters provide the tools needed to examine manure applications and their effects on the environment. If manure is distributed on cropland, pastures, or other lands without considering the capacity of the soils and vegetation to use and store the nutrients in the manure, these nutrients can be over applied and available for transport off site. When this occurs, manure provides a source of pollutants for waterways and aquifers. An additional pathway for nitrogen in manure is through **volatilization** as NH_3 or N_2O. Carbon is also emitted both as CO_2 and as methane (CH_4). Based on the storage and atmospheric conditions, these gases can be deposited miles away attached to particulates, fall contained in precipitation, or add to atmospheric greenhouse gases (Aillery *et al.*, 2005; Letson & Gollehon, 2007).

What is waste?

The dictionary definition of waste is whatever we humans do not want. This makes sense to us except that this definition does not explain why waste is a negative, that is it is bad for the environment. It provides no test for what is simply a nuisance and what, if not properly managed, can cause serious harm to us or to the integrity of our environment.

This definition is simply not adequate for natural resource management or protection of the environment. A useful, scientific, definition is one that is not tied to personal preferences but is instead directly linked to the impact an object has on its surroundings and how it functions. The term needs to indicate what can cause ecosystem imbalances.

If we are to achieve a degree of sustainability in our management of our air, water, and soil resources, we have to examine what is being added to the resource from its perspective.

5.6.1.6 *Fertilizers and agricultural chemicals*

Crop production is greatly enhanced by the application of nitrogen, phosphorus, potassium, and other fertilizers to boast plant growth and the use of pesticides to reduce losses to insects, disease, and other pests. However, crops are not 100% efficient in using these substances, which means they are available to leave the site, either in surface runoff or, if soluble, in subsurface flows. Because there is approximately 158 million hectares (390 million acres) of cropland in the United States (Natural Resource Conservation Service, 2018) and over 13.4 billion hectares (33.1 billion acres) of cropland in the world (Food and Agriculture Organization, 2003), even small amounts of these substances leaving individual fields can, in aggregate, result in substantial additions of nitrogen, phosphorus, and other chemicals entering waterways.

In agricultural operations, nutrients can be applied as either solids or liquids. It is easy to see how liquids might be transported by water either in surface or subsurface flows, but solids need to shift phases to be transported in subsurface water flow. We are all familiar with compounds shifting phases. It is what we observe when we add sugar to coffee or iced tea. What happens? The granules dissolve into the fluid and apparently disappear. We of course know better and can attest to the sugar's presence by taking a sip of our drink. This is sugar shifting from a solid phase by entering into solution. Soluble forms of nutrients such as nitrate and to a lesser extent phosphate do the same thing. Unlike our drink it would be unwise to test for dissolved nutrients or other chemicals by drinking water from irrigation ditches or streams. However, we can be reasonably certain that if soluble compounds are present and water is available to transport them to waterways, that these compounds will be present in the water. That said, without conducting tests on this water we will not know how much of these compounds is being moved into waterways.

> **Experiment 5.1:** Dissolve a teaspoon of sugar into an 8-ounce glass of room temperature water – what happens? Dissolve colored sugar from a pixie stick in water – what happens? Now return to the first glass of water, add additional sugar into the water one spoonful at a time. What happens as you add more and more sugar to the glass? Now gradually warm the water what happens? Cool the glass. What happens?

This experiment demonstrated that there comes a **saturation** point when water can hold no more of a compound and it precipitates. At this point, the available water will not be able to transport additional amounts of the dissolved compound although the water can transport the compound in its solid form. This is an important factor to consider with respect to phosphorus compounds which are not as soluble as nitrogen compounds.

As we discussed in Chapter 3, numerous forms of nitrogen (e.g. NO_2, NO_3, and NH_4) and phosphorus (P_2O_5) are water soluble and susceptible to water transport. The same is true for numerous **herbicides**, **insecticides**, **fungicides**, and other chemicals in common use in modern agricultural production systems.

5.6.1.7 Other sources of nutrients and carbon

Energy production, industry, and transportation are sources of nitrogen and carbon. These sectors release nitrogen into the atmosphere that deposits, generally but not exclusively with precipitation, on land and waters contributing to the nitrogen cascade discussed in Chapter 3. Hydrocarbon mining and processing results in the inadvertent release of methane and other gases, and the production of uncaptured byproducts some of which escape into the environment. These gases and byproducts contribute nitrogen and carbon into the atmosphere, waterways, and industrial waste disposal areas. Industrial processes often use hydrocarbon energy sources that release carbon and nitrogen into the atmosphere and waterways and generate byproducts. Similarly, transportation systems largely rely on vehicle engines that rely on hydrocarbon combustion, which in aggregate release large amounts of carbon dioxide and nitrogen oxides into the atmosphere. To be effective environmental management strategies need to account for the carbon, nutrients, and other pollutants contribution of these sectors into ecosystems.

Chapter 6

Impacts of human-caused changes to water flow and to the balancing of the carbon, nitrogen, and phosphorus cycles

Nature sides with the hidden flaw.

—Murphy's Laws

In the previous chapter, we discussed how humans have altered natural systems, generally to their benefit, or at least to the perceived short-term benefit. We showed how the components of our environment relate to each other in a balance among cycles moving from one key carbon, nitrogen, phosphorus (CNP) pool to the next. Human population growth has demanded tremendous modifications of natural landscapes and shifts in how resources are used and where they go. Natural systems undisturbed by us could not have sustained this population growth. In this chapter, we introduce you to the challenges in protecting and managing our natural resources that we now face as a consequence of these shifts. Wise management of these resources is necessary if we are to continue to enjoy their benefits.

We first identify several categories of natural resource and environmental impacts and then help you make the connection between how altering water stocks and flows and disturbing the balance of the carbon, nitrogen, and phosphorus cycles contributed to these impacts. The four main categories of impacts we discuss are floods and droughts; diminution or impairment of water quality, with associated impacts on the quality of both aquatic and terrestrial habitats; diminished air quality; and climate change. The latter relates to all of the above, just as all of the above are also interconnected components of a larger system, although on different timescales. The discussion illustrates how an action leads to an effect. At this point it should be clear to you why these impacts are 'sticky' and why addressing the negative impacts is so difficult.

We emphasize that in our discussion of water movement we are referring to freshwater resources. It is important to bear this in mind because although 71% of the world's surface is water, only 2.5% of water is freshwater, its relative scarcity makes it critically important for human welfare (Bureau of Reclamation, 2020). As populations grow, useable water for drinking and agriculture is expected to become a constraining resource (Gleich, 2017; WHO, 2019). This limitation has already been observed in Sub Saharan Africa, South Africa, the western United States, and elsewhere. Actions that can conserve clean, freshwater will become more important.

6.1 IMPACTS FROM FLOODS AND DROUGHTS

The number and intensity of floods and droughts in the United States, including associated wind-related events, have increased over the past forty years with concomitant damages increasing as well (U.S. Global Change Research Program, 2018a). Figure 6.1a shows the number of flood and drought-related events since 1980, with Figure 6.1b showing the estimated damages per year associated with the events. What is clear is that the upward trend for damages is beyond the expected year to year variation. What cannot be discerned from these data is the percentage of the damage cost that could have been avoided through better planning, a topic we discuss in Chapter 7.

6.1.1 Flood damages

Floods are part of the natural cycle of weather events and water flows. However, flood damage is a human concept. Early farmers recognized that annual spring

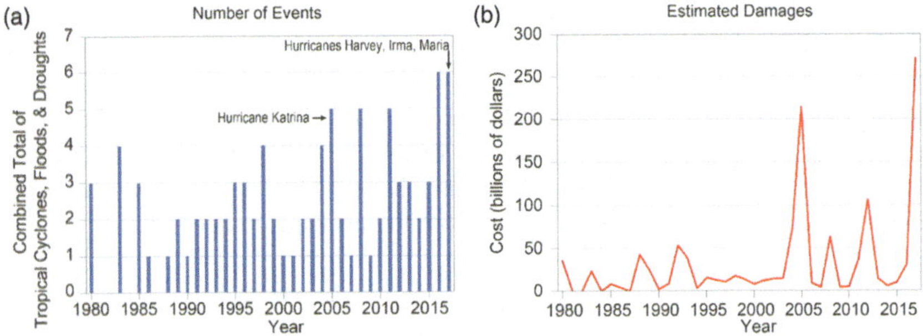

Figure 6.1 (a and b) Number of flood events and the associated damages. *Source*: U.S. Global Change Research Program (2018a).

floods brought nutrients to their cropland, increasing yields. Their cultures adapted to these floods, and generally viewed them as beneficial. However, flooding also killed livestock, ruined crops, damaged dwellings, and impeded river traffic. The extent of damage that is caused depends on what is flooded, flood level and duration, and what is in the water. Ironically, because floodplains are flat, are located conveniently for river transportation, and often have fertile soils, they are attractive sites for establishing towns, cites, and farms. As settlements expanded and more structures were constructed, more flood damage occurred.

One can classify the nature of flood damages by (1) damages associated with natural variation in water levels resulting from having placed crops, livestock, and structures in the wrong place and (2) damages to livelihoods and property resulting from increased water levels resulting from human alteration of natural water flow. The latter is of more interest to us here even though the former too often occurs as a result of human hubris, ignorance, or willingness to gamble with the probability of a flood event. In the former category, we put the unwise construction of homes in drained wetlands or floodplains unprotected or inadequately protected by levees. Levees provide protection in the event of a high water flow event provided that the water does not breach the levee or exceed its height. Levees constructed by other communities can cause water levels to rise for other communities located on the same river. These rising waters can overtop levees built to handle 'natural' flood levels.

As we have repeated throughout this text, water, or anything else for that matter, does not simply disappear. It goes somewhere and furthermore it goes downhill. And the speed of water matters. Human actions change the speed and pathways of water movement. Altering drainage patterns with barriers, drainage systems, or reservoirs changes how fast water moves and how quickly water arrives in waterways. Changing vegetative covers alters infiltration and evapotranspiration levels which change the proportion of water that moves into the atmosphere versus surface flow. Human intervention includes vegetative change, irrigation and drainage systems, levees, and dams (Chapter 5).

Constructing levees to protect human activities from flooding alters the natural water flows. Conducting activities in floodplains and in areas susceptible to flood damage increases the potential for damage. [Less intensive use, such as leaving the land in parks, pastures, and forests mitigates this risk.] Any upstream activity or modification to the landscape that increases runoff can contribute to the risk of flood damage, with the possible exception of major rainfall events.

Storm surges associated with hurricanes can also cause flooding. Even though they are natural phenomena, their timing or magnitude cannot, at this time, be precisely anticipated. The severity of their impacts can be exacerbated by human actions that can increase their damage, such as degrading or removing coastal wetlands and barrier islands that otherwise would diminish the energy and volume of water associated with these events.

6.1.2 Drought impacts

In the United States the economic costs of droughts amount to some $9 billion per year, generally to agricultural interests (NOAA, 2020). The damages result from a variety of economic insults, including reduced water for drinking, irrigation, and industrial use. River base flow below the depth necessary for barges and other vessels impedes river transportation. Delays increase costs and can cause a diminution of the value of the goods transported. In watersheds where human interventions have reduced subsurface base flow, the likelihood for low flows sufficient to adversely impact human activity is increased. The impact of low flow can be particularly acute when the rights to withdraw water from the steam have been based on stream level from years with high flow.

Droughts may appear to be solely a consequence of climatic variation. However, their severity and frequency are affected by human actions. Land management affects the vulnerability of the soil and the vegetation to periods of low precipitation. Humans exacerbate the impacts of these droughts by how they manage the organic matter in the soil and the vegetation. Even in river systems where extreme deficits of precipitation have not occurred, humans can create situations of water scarcity, generally downstream of dams where water flows can be regulated, by diverting water for irrigation or other uses thereby depriving downstream users of the water. Redirecting water from rivers further serves to reduce base flow, depriving downstream habitats of water raising the potential for ecological damage.

Altered land use in the form of urbanization has increased impervious surface area, leading to more rapid runoff and reduced infiltration, which in turn reduces longer term water storage. Our need to grow crops for food has altered the land cover over vast expanses of arable land. In the United States of the approximately 802 million hectares (3.1 million square miles) of land area, approximately 28% has been altered by humans for use as cultivated cropland, pastures (22%) or settlements (6%) (Sleeter *et al.*, 2018). Changing the land cover by converting native grass or forest land to crop use, alters the flow of energy, water, and greenhouse gases from the land to the atmosphere. Agricultural lands tend to have more runoff and less soil water storage than land with native vegetation. Also, agricultural uses often require drainage or irrigation systems, both of which decrease water storage. In droughts, the reduced water storage associated with urban and agricultural land increases the drought's impacts.

The organic component of soils has declined precipitously since the mid-nineteenth-century with the expansion of agriculture. Degrading soils along with **deforestation** and conversion of grassland to cropland was a major source of carbon to the atmosphere (Lal *et al.*, 1997). In the agricultural breadbasket of the United States, roughly half the organic matter in cropland soils has been lost, largely as a consequence of tillage practices that exposed the soil to oxidation (Natural Resource Conservation Service, 2013). The soil organic matter holds much of the water that sustains growing plants in periods of low precipitation. Studies have shown with a 1% increase in soil organic matter, the available water holding capacity in the soil increased by 3.7% (Food and Agricultural Organization, 2003). The loss of one half of the 6% carbon from these soils initially contained results in a more than 10% loss of the capacity to store water.

6.2 WATER QUALITY IMPACTS

Rivers and streams, as explained in Chapter 2, transport sediment and nutrients. Human activity has increased the quantity of both sediment and nutrients transported by rivers. According to United States Environmental Protection Agency, as shown in Figure 6.2, the biological condition of waterways in the United States is generally poor. Some half of the streams and rivers are in poor condition and another quarter in fair condition, as assessed against biological indicators. Nitrogen and phosphorus predominate as chemical stressors, with 41% of the surveyed waterway reaches impaired, that is fail to meet water quality standards for their designated use, by nitrogen and 46% by phosphorus. Acidification accounts for 1%. The survey found that 24% of the rivers and 20% of the streams suffered from high levels of riparian disturbance – that is, human activity. Fifteen percent of rivers and streams exhibited excess streambed sedimentation.

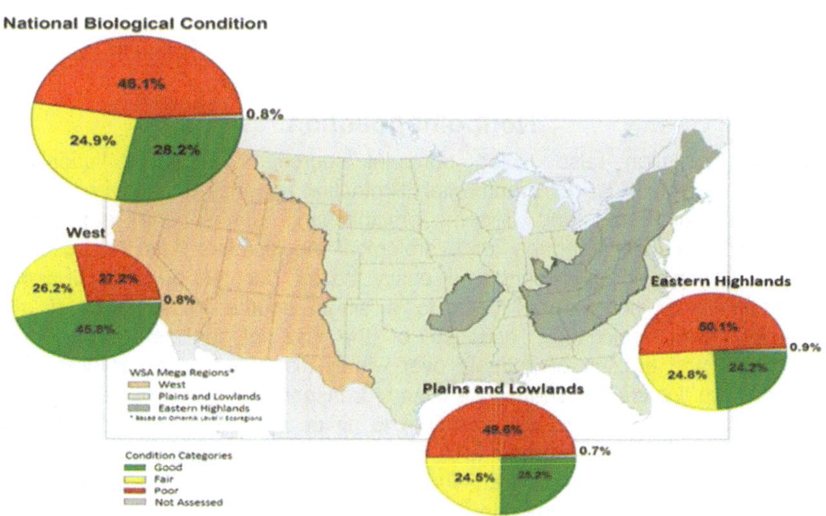

Figure 6.2 Biological condition of the nation's rivers and streams, based on the Macroinvertebrate Multi-metric Index. Adapted from U.S. Environmental Protection Agency (2016).

These findings should be disturbing to anyone concerned about the quality of our environment. Keep in mind, however, that the conditions of waterways in general have actually improved since the major environmental laws in the United States went into effect in the 1970s. Progress has been made with emissions from urban centers – smokestacks and sewage and industrial drainpipes (U.S. Environmental Protection Agency, 2020a). In most of these cases, a control technology could be identified and installed to reduce pollution. As this book has provided you with the tools to evaluate problems, you can surmise that progress has stalled in protecting rivers and streams from nonpoint pollution (that is to say sources other than smokestacks and drainpipes). These sticky problems remain.

Streams flow into rivers and rivers flow into estuaries and near coastal waters and ultimately the oceans. What can we say about the conditions of estuaries and the oceans? According to the U.S. Environmental Protection Agency, some 30% of bays and estuaries are impaired. Major stressors are nutrients and sediment. Pollution transported through the air and deposited to water bodies is also a significant threat. Sediment dispersed in water makes waterbodies murky reducing the amount of solar radiation that reaches the underwater (subaquatic) vegetation. Eutrophication, the enrichment of water with dissolved nutrients, leads to a dense growth of microbial and plant life that suffocates fish and other animals in oxygen-poor (hypoxic) water. And what is happening to the oceans? An important impact is **dead zones (hypoxia)** from nutrient pollution – the nutrients that entered the waterbodies causing environmental effects discussed above eventually make their way to the ocean where they cause further damage.

Nonpoint pollution

Nonpoint pollution (also called nonpoint source or diffuse pollution) is defined by the U.S. Environmental Protection Agency as environmental contamination from land runoff, precipitation, atmospheric deposition, dispersed drainage, seepage, or hydrological modification. It is the pollution that is transported but cannot be easily traced (for the purpose of legal enforcement or compliance) back to a single source. Most (but not all) pollution from agriculture, because of the use of fertilizers and other chemicals and grazing of animals over widely dispersed areas, is classified as nonpoint. Where the pollution is clearly transported by identifiable conduits (such as a pipe attached to a building) to air or water it may be defined as point source pollution, depending upon the size of the operation and other factors. Pollution does not necessarily have to be chemical or biological in nature. An activity that results in the elevation of the temperature of a stream or river to levels that is detrimental to the viability of the ecosystem is also defined as pollution.

Figure 6.3 Hypoxic zones around the world. Red circles represent areas of hypoxia with intensity reflected by the size of the circles. *Source*: National Space and Atmospheric Administration (2008).

Figure 6.3 shows the location of coastal and ocean dead zones throughout the continents and oceans, adapted from the National Space and Atmospheric Administration (2008) earth observatory. The size of the dots corresponds to the area of the dead zone. Though the data are from 2008, the message remains the same for 2020, just more serious. In the 1960s, scientists had identified 49 dead zones. As of 2005, there were 405, according to the study conducted at that time (Biello, 2008).

6.2.1 Environmental impacts of soil erosion and sediment transport

Soil, nutrients, and other materials leaving a site have an impact when they show up elsewhere. These effects are referred to as offsite or downstream impacts. They are different from damages from reduced productivity, soil carbon, or the loss of nutrient availability because these impacts are borne by individuals downstream or downwind from the operation rather than the operator.

6.2.1.1 How water erodes soils

Soil erosion, the displacement and transport of soil, is the bane of agriculture. Soil erosion takes place with or without human intervention, but human activities, such as altering the native vegetation (this includes land conversion to cropland), exposing and disturbing soils, and creating impervious surfaces, lead to accelerated rates of erosion. Tilling the soil is an example of soil disturbance and is a major cause of erosion from croplands. A direct economic consequence of erosion results is the offsite damages downstream resulting from degraded water quality and the cost of dealing with soil deposited in ditches, reservoirs, channels, and harbors. When this accelerated erosion occurs on cropland it reduces agricultural productivity. For the remainder of this section we will refer to accelerated soil erosion due to human activity as either erosion or soil erosion for convenience.

Water is the primary agent for soil erosion. The force of water on soil moves billions of tons of sediment each year. In 2015, USDA's Natural Resource Conservation Service (NRCS) estimated that water caused an average of 6.1 metric tons of soil erosion per hectare (2.71 tons per acre) of United States cropland, resulting in an estimated 0.9 billion metric tons of erosion (0.99 billion tons, U.S. units) (Natural Resources Conservation Service, 2018). This is a 34% decrease from 1.45 billion metric tons per year (1.60 billion tons) in 1982.

Soil erosion is a good starting place for examining how natural forces combine with human activities to have environmental consequences and to illustrate the point that gradual change can be overlooked. By now you are most likely tired of us repeating the basic points – everything goes somewhere, gravity does not take a vacation, observe what is happening, and think things through. That said examining erosion demonstrates how these tools can provide a better understanding of natural processes and how human intervention can either exacerbate or mitigate conditions.

Soil erosion is also a good place to demonstrate that measurement is an important tool by illustrating that gradual changes can be overlooked. Consider the rate of erosion noted in the previous paragraph, 6.1 metric tons (2.71 U.S. tons) of soil eroded per cropland hectare (acre). This is over 2,230 kilograms (5,420 pounds) of soil eroding on each hectare (acre) (approximately 610 grams per square meter or 2 ounces per square foot), which is hard to observe without measurement because the erosion on most cropland is gradual and mostly uniform across the field. This form of erosion is referred to as sheet erosion. Rill erosion, which often occurs in conjunction sheet erosion, is when the erosion is slightly concentrated in shallow channels (rills) that can be observed in a field. It was not until U.S. conservationist Hugh Hammond Bennett conducted his ground-breaking work around the turn of the twentieth century, that sheet and rill erosion was recognized as a threat to agricultural productivity. The offsite damages are due to the deposition of sediment increasing the cost of maintaining drainage and irrigation ditches, dredging channels and harbors, filtering water for drinking and industrial uses, shorter useful life spans for reservoirs and hydroelectric plants, and reduced recreational benefits.

Experiment 6.1: Take a pail of soil and dump it on a sloping ground where it is exposed to rainfall. Measure its height, protect it from foot traffic, and observe it over time. Why does the height decrease? Does the volume decrease? If so where does the soil go?

Soil erosion can be viewed as a function with four components, the characteristics of the landscape (particularly topography), water energy against soil particles, soil properties, and the disturbance and exposure of the soils to erosive forces. Wind is also an erosive force that is important in more arid areas with substantial wind and exposed soils. We are going to focus on water erosion, but encourage the reader to consider how wind erosion might be similar and different from soil erosion. Two key concepts to consider are the different properties of wind and water and differences in their movement. We have already discussed soil properties and exposure in Chapter 4. We will now examine how the landscape and water energy affect soil erosion and then look at soil conservation, actions that can be undertaken to reduce erosion.

Landscape: The topography plays an important role in erosion by influencing the amount of runoff and its velocity. The geologic characteristics of the land and human modifications will affect water movement. Large amounts of impervious surfaces such as rocks and pavement within a landscape reduce water infiltration into soils, increasing surface runoff. The slope and length of the slopes within the overall landscape and individual fields affect the concentration and velocity of water. A steep, narrow watershed will funnel water quickly into a concentrated flow. In contrast, a broad, flat watershed characterized by healthy soils will have less surface runoff per unit areas and have slower flow rates. The faster the water moves, the greater the energy it has to erode soils.

A **Karst** geology is a feature of some regions that can be locally important for water movement. Karst landscapes are characterized by caves and sinkholes created from water chemically eroding limestone deposits. These caves and sinkholes are a pathway to subterranean channels that water enters during and after storm events. Because they divert water from surface runoff, they can reduce soil erosion. Karst can rapidly transport the water, along with materials dissolved in water, to rivers and streams.

Energy: It is clear that as more force is applied to an object or objects, more material will be moved a greater distance. The amount of force water exerts is a function of its mass (think about the amount of water) and velocity. Climate has a large role on the amount of erosion because most water coming into contact with the soil comes from precipitation. The force of water in the form of rainfall can break up soil aggregates and dislodge soil particles. Rainfall is far more important than snow in determining the amount of erosion because falling snow is less dense and falls more slowly than rain transferring less energy to soils. When rain falls, particularly on bare soils, the raindrops transmit energy against soil particles. This can be observed by looking at the first raindrops falling on a dusty soil. The splash that you can observe is somewhat like a miniature meteor impact or throwing a rock into a pond. And like throwing rocks into a pond, the amount elevated and displaced increases with the size of the drop and the quantity of drops. If you look carefully at rain falling on a dusty soil, you can see the raindrops displace and elevate soil particles, which land in a ring around the point of impact. If you do not want to wait for a storm or stand out in the rain, you can use an eyedropper and a pail of soil. Driplines, depressions in the soil where water falls from rooves or other structures, provide additional evidence that rain moves soil.

Snow melting can influence erosion, particularly in years with high snow accumulation, by storing water for release in the spring. When the snow melt is associated with substantial rainfall, the snow can melt quickly releasing additional water substantially increasing runoff from the rainfall. This can result in flooding. Snow, ice, and frozen soils can also reduce infiltration increasing surface runoff. Frozen soils also can bind soil particles together protecting them from erosion. However, as the soils thaw, they become vulnerable to erosion (Sharratt *et al.*, 2000; Zuzel *et al.*, 1982).

Irrigation and flooding also bring water into contact with soil. Some forms of irrigation can result in substantial erosion. The effect of flooding on erosion depends on whether the velocity of water is sufficient to transport soil particles or is slowing allowing sediment to settle. Where flood waters are moving rapidly, they scour the landscape causing large amounts of erosion. Where flood waters slow into in large portions of the broader floodplain, sediment in the water is deposited. An area can both be scoured and have soil deposited during the same flood event.

(**Discussion Question:** How do you think different sized soil particles will be distributed within a floodplain?)

Areas with high rainfall will generally have greater amounts of water-based erosion than areas with relatively low amounts of rainfall. However, the distribution of rainfall events and intensity of the rainfall is also important. All other things being equal, for two regions with the same annual precipitation, one with an even annual rainfall distribution and the other characterized by wet and dry seasons, the region with wet periods will have more erosion. This is because the seasonal rainfall will lead to greater volumes of water in a shorter period of time, concentrating more force against soils (Mohamadi & Kavian, 2015). Similarly, areas with intense rainfall events will have more erosion than one with the same amount of rain falling over a longer period. Studies have estimated that the years with the greatest rainfall have substantially more land with high rates of water erosion (Conservation Effects Assessment Project, 2012).

Experiment 6.2: Observe the soil below a gutter downspout. Look for one where the water flows onto bare soil. Note the rills or small gully that can result from concentrated water flow. Now compare the speed of the water flowing from the gutter with that of sheet flow on the adjoining pavement. This is an example of the increased force from a concentrated flow of water versus that of a more dispersed flow.

> **Experiment 6.3:** Take a hose and direct the spray against bare soil alternating the flow from the nozzle. Note that as the flow becomes more concentrated, more soil is dislodged.

Soil conservation: Maintaining soil productivity is essential for agriculture. Agriculture has a large footprint on the landscape, and soil erosion in most cases reduces soil productivity. For these reasons, soil conservation has traditionally focused on agricultural soils. Conservation practices have been developed and tested by soil scientists and farmers to protect these soils. They examined the factors that contribute to soils being more vulnerable to erosion to identify factors that could be changed to reduce erosion. Clearly, the broad landscape, soil texture, and precipitation are not things that can be readily changed. On the other hand, reducing energy from overland water flow, adopting in-field conservation practices to increase infiltration, increasing protection from vegetative covers, and modifying production systems to reduce soil exposure are practices that can be adopted relatively quickly. Soil structure and organic matter content can be changed, but these changes require a long-term conservation strategy and a multi-year time period.

Actions that can reduce energy from overland flow include reducing the length of the slope, reducing the slope, and reducing the amount of water flowing overland. At first blush, it might not seem simple to reduce the length of a slope, but recognize that placing a small barrier on your driveway will slow water movement. This is because it reduces the length water can flow unimpeded. In agriculture, similar obstructions can be placed along a slope which act to reduce the slope length, slow water movement, and increase infiltration to reduce erosion. These practices range from contour tillage and **contour farming** which arrange crop rows across the slope rather than up and down the slope, to constructing terraces which act as steps along the slope. These practices serve to reduce the speed of water moving across the field, by in effect reducing the slope length. Contour tillage and farming do this with tillage and crop orientation, while terraces do this by reconfiguring the landscape. Another approach is strip cropping where two crops, one of which provides greater soil protection and slows water flow, are planted across the slope.

> **Observation:** Have you ever walked on a trail where wood or stone strips periodically crossed the path? These strips were placed to reduce erosion by reducing the slope length for water.

Vegetative covers provide a tool to reduce the exposure of soils to rainfall and surface flow. Year-round covers that have crown closure provide the greatest protection from rain splash. Permanent or semi-permanent cover such as those with closely grown stems such as hay or grass slow water movement and provide the greatest resistance from the force from overland flow. Vegetative covers such as these might be employed on soils and sites most vulnerable to erosion. They are not feasible for less vulnerable, highly productive cropland. On these sites the crop production systems can incorporate: vegetative management practices, such as conservation **crop rotations**, no-till or reduced tillage; harvesting practices that assure adequate crop residues; and/or **cover crops** to protect the soil when crop foliage is not adequate. Appropriate production systems are developed by agronomists or conservation specialists using site-specific information on the farm, the soils, the climate, economics affecting the farm, and other key factors.

Let us examine more closely how crop rotations, **tillage** and harvest operations, and cover crops can protect soils from erosion. Crop rotations, the sequence of crops planned for a field, determine numerous factors for crop production including expected planting and harvest dates, soil exposure to the elements, nutrient requirements, pathogen and insect cycles, and field operations. Here, we will focus on the planned vegetative cover for the growing season for each of the years within the rotation, but need to emphasize that that crop rotations are about more than soil erosion. That said, because for a given location each crop within the rotation has an expected planting date, crop development profile, harvest date, and expected plant residue after harvest, choosing an appropriate crop rotation is an important first step in reducing soil exposure and thereby protecting soils against erosion. For example, crops planted earlier in the season and having more rapid crown closure can provide greater soil protection. Additionally, crop rotations that increase soil organic matter will enhance soil productivity, structure, and health, increasing infiltration over the long run. Tillage systems that reduce soil disturbance protect soils from erosion by minimizing soil exposure, maintaining soil structure, and reducing oxidation of soil carbon. Retaining soil carbon helps maintain soil structure which helps water infiltration and enables soils to better withstand erosive forces. Harvest systems that maintain vegetative residue also reduce erosion by providing a protective cover on soil surfaces. The residue also provides organic matter that is a source of future soil carbon. Increased infiltration, higher soil carbon, and good soil structure all help reduce soil erosion.

Cover crops are grown to protect soils outside of the primary growing season. These crops are not the primary cash crop, but act as protective vegetative cover after the harvest of the primary crop, provide organic matter as a source for soil carbon, and depending on the crop might generate income, be grazed, fix nitrogen, or serve another function. They are also beneficial in capturing soluble nutrients, holding them on site. Cover crops can be established as the crop is harvested or when the field is prepared for the following growing season. Under some production systems crops can be planted directly into some cover crops which can then protect the soil until the cash crop reaches crown closure (Tilley & Winger, 2014; Blanco-Canqui *et al.*, 2015).

Drainage systems also reduce soil erosion by reducing overland flow, but the increased water infiltration has other consequences that need be discussed later.

Drainage systems are installed in fields where the high water levels inhibit crop root growth. These systems permit naturally saturated soils to drain to a depth that can range from 3 to 6 feet (0.9–1.8 meters) permitting adequate root growth. Drainage moves water rapidly through either subsurface tile drains or drainage ditches directly into waterways. These systems allow water to drain from soil pores, creating open pore space for infiltration during rainfall events, reducing the amount of surface runoff and erosion.

6.2.1.2 Downstream impacts of soil erosion

Soil leaving a field is carried by runoff and is either deposited on land downhill from where it started or into a body of water. The sediment component deposited on land might be on a field where it has little impact, or in drainage ditch, irrigation system, roadway, or other infrastructure where it needs to be removed to maintain the function of the structure. The cost of removing the sediment is an offsite cost from soil erosion. Additional terrestrial-based offsite costs occur when sediment fills in wetlands reducing or eliminating the benefits these wetlands provide (e.g., **nutrient cycling**, water retention, wildlife habitat, and carbon sequestration). Pimentel *et al.* (1995) estimated total water erosion offsite cost to be $12.6 billion annually.

Sediment going into water bodies can be suspended in the water column or deposited in rivers, streams, lakes, reservoirs, bays, and oceans, reducing water quality, decreasing storage capacity for reservoirs and irrigation ponds, obstructing navigation, diminishing recreational value, and reducing fish and wildlife habitat. These impacts are referred to as instream effects. The cost of these instream impacts has been estimated for numerous locations (Telles *et al.*, 2011).

Reduced water quality from suspended sediment has several components: increased turbidity, higher water temperatures, elevated nitrogen and phosphorus levels contributing to **eutrophication**, and degraded fish habitat. The sediment intercepts light which reduces the light available for algae and other aquatic vegetation and raises the water temperature. The suspected sediment degrades aquatic habitat by harming fish gills, inhibiting mollusk feeding and respiration, and lowering visibility which reduces fish mobility and hunting foraging success.

Sediment suspended in the water column is eventually deposited either within water bodies or on floodplains and has been estimated to cause between $12.6 and $18 billion of damage in the United States (Clark, 1985; Ribaudo, 1986; Pimentel *et al.*, 1995; Telles *et al.*, 2011). These effects take place along the full length of the waterbody. Sediment deposited in pools within upstream reaches can fill important fish habitat and bury important invertebrate habitat. Sediment deposited further downstream can form sand bars that block navigation, shift channels, or fill up important ports. Impacts on navigation include increased travel time due to constrained and shifting channels, diminished port use, and substantial dredging costs. These costs have been estimated to be between $365 and $1,508 million annually (Ribaudo, 1986; Hansen *et al.*, 2002).

Sediment deposited in reservoirs reduces the lifespan of the reservoir and storage capacity. Because a reservoir's usefulness for flood control, energy generation, and irrigation and municipal water supply is a function of its storage capacity, accelerated sedimentation has significant costs. Lakes and reservoirs are important recreational sites for fishing, boating, and related activities. Reducing their useful life reduces recreation opportunities. Ribaudo (1986) estimated the erosion costs from lost reservoir storage capacity and associated maintenance to be $2.4 billion annually. Hansen and Hellerstein (2007) estimated soil conservation efforts between 1982 and 1997, provided a $223 million annual benefit in 1997 due to preserved storage in reservoirs.

6.2.1.3 Wind erosion

Wind erosion moves soil, reducing productivity where it occurs, causing damage from diminished air quality, increased maintenance and cleaning costs, crop injury, and health impacts. Wind erosion is of a greater concern in more arid regions such as the western United States, but does occur to a lesser extent in humid areas. Because wind, unlike rivers and streams, is not constrained within channels, the damage from wind erosion is more dispersed, and does not concentrate sediment and nutrients in the same manner as water erosion. Factors associated with wind erosion are wind velocity, dry soils from arid climates, soil exposure, and soil characteristics. Wind erosion can involve particles bouncing along the ground or suspended in the air. Both of these forms reduce soil productivity and can damage crops, but particles suspended in the air cause greater damage (Huszar & Piper, 1986; Piper, 1989).

Conservation practices can reduce wind erosion. These practices focus on reducing the velocity of the wind at the soil surface and reducing soil exposure. Establishing a wall of vegetation perpendicular to the prevailing winds slows wind speeds at the land surface. Windrows of trees, hedges, or grass can force air movement upward creating a wind shadow behind the row. When appropriately spaced, these windrows reduce erosion by reducing wind speed at the soil surface. Practices that reduce soil exposure to wind include establishing and maintaining vegetative covers, increasing crop residues after harvest, and adopting tillage practices that maintain crop residue on the soil surface. These practices reduce both the force of the wind on the soil surface and soil exposure, and therefore soil erosion.

6.2.2 Impacts of added nutrients

Modern agriculture requires the application of nutrients (natural or artificial) to achieve yields necessary to feed a growing population and economic success. For many lands, fertilizer is essential for compensating for eroded soils and degraded environments resulting from decades of poor management of soil resources (see Dust Bowl). The application of fertilizer and manure involves a tradeoff between food and financial security and the environment. A farmer's best efforts will not have all of the fertilizer he applies taken up by his crops. Some will escape into the atmosphere and waterways. Farmers face a balancing act. Apply too little fertilizer and the crop yield is subpar; too much and the excess becomes available to be transported away. Given that no one can predict the weather, often farmers tend to err on the plus side, as a form of insurance policy (Canfield, *et al.*, 2010; Robertson *et al.*, 2013; Smil, 2001). In the case of corn, some have estimated that the excess is about 50% of what was applied (Canfield, *et al.*, 2010; Robertson *et al.*, 2013; Smil, 2077).

Following nutrients and tracing their impact as they move downstream is more complicated than for sediment. Excess nutrients introduced into the water column can be taken up by plants and microorganisms or transformed into different compounds as they are transported downstream. The multiple pathways nutrients can take require careful examination to assure we have fully accounted for all nutrient components introduced into the system. Identifying the consequences of increased nutrients leaving the land requires examination of nutrients in the water column and the downstream impacts of lower water quality, plant uptake, volatilization, and denitrification. Complicating matter further is the capacity of nutrients to be transported and deposited as sediment on floodplains and in stream banks and channels, only to be released later when the sediment is eroded again. These legacy deposits are particularly important for phosphorus.

Nutrients, especially nitrogen can transform into a variety of molecules, each with its own environmental impact and potential to change media, from soil to water to air. The consequence is that the constant introduction of additional rN into the biosphere creates a dynamic disequilibrium whereby each year there is more and more excess rN available, disrupting ecosystems.

Figure 6.4 illustrates the dilemma. Without human interference, nature balances the rN (as well as bioavailable phosphorus although we do not show this here), whereby the reactive material that is generated by natural processes like lightening and leguminous plants is roughly equaled by the reactive material that

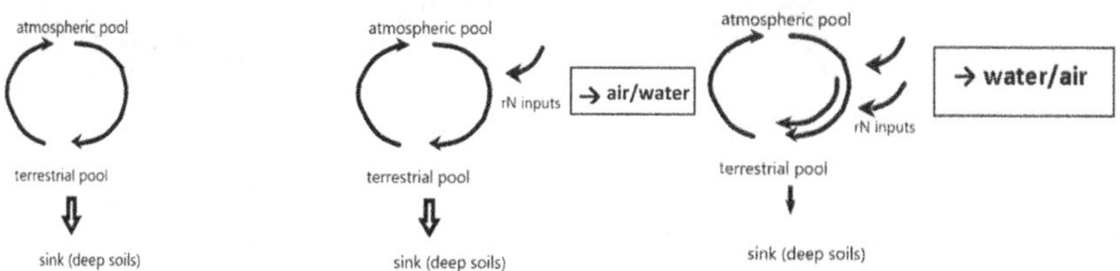

Figure 6.4 Reactive nitrogen: natural condition, with additional rN inputs and with additional rN over time.

is returned to the atmosphere by denitrification (left arrow) or stored in soil or oceans (bottom arrow). The nitrogen stored in soils and the ocean is no longer available for reactions in the biosphere. The middle figure shows more rN from fertilizer and nitrogen-fixing plants added into this cycle. The system must readjust to accommodate this additional reactive material, but in the short term, there is an excess of material that escapes into the environment. This material pollutes the water and atmosphere and disrupts ecosystems. The figure on the right shows the consequence of repeated years of the addition of new reactive material. Not only does the system have to adjust to the material added the time period before, but it must now accommodate the additional infusions. More excess reactive material builds up in the system and the flows between the pools increase. Moreover, landscapes have been modified and soils tilled so that the amount of reactive material captured in deep storage decreases. Also, wetlands have been converted or drained, ending their capacity to denitrify rN, reducing the return of rN back into the atmosphere. We are now introducing 50% more rN into the environment than nature does. The balance that once existed cannot be achieved until equilibrium between extraneous rN inputs and the sum of denitrification and long-term rN removal into sinks is reestablished. In lieu of this, the ever-growing amount of reactive material pollutes water and air, disrupting ecosystems and the services they provide. How this is so is explained below.

Nutrients, particularly reactive nitrogen, act differently from what many of us think of pollutants. The common perception is that pollutants leave the source, cause disruption, and then dissipate or break down in the larger environment (Figure 6.5a). Nutrients, because of their very basic nature, do not behave in this manner. They stick around in the environment until they are taken up by plants, destroyed and converted back to an inert or nonbioavailable form, or go into long-term storage, such as in soils. They do not just pollute once, as we show in

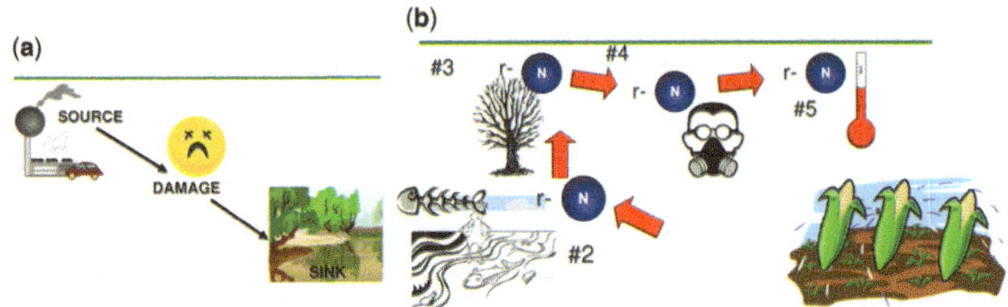

Figure 6.5 (a) Typical pollutants enter the environment, do their damage, and then leave through a sink that demobilizes them. (b) Serial offender pollutants. After being applied to crop as fertilizer rN travels the environment like a pinball. In water #2, as nitrate, it can kill fish. In air #3 as ammonia, it can damage ecosystems #4 and kill animals. And in the atmosphere #5 as nitrous oxide, it acts as a greenhouse gas.

Figure 6.5b, but have a cascade effect. The same molecule transforms causing serial damage. When we apply rN as a fertilizer, only a portion is actually taken up by the plant fertilized, while the rest is lost to the environment. The proportion lost to the environment increases with increasing precipitation and temperature. In the case of corn, more than half can remain in the soil profile, some of which can be used to build new soil, as was explained in Chapter 4 (Sawyer, 2014). But depending how the land is managed, the residual nitrogen can be transported by water to a stream or pond or volatilized into the atmosphere, where it can cause damage. The reactive nitrogen can transform repeatedly; into ammonia, where it can damage vegetation, into NOx, an air pollutant that can negatively affect air quality and human health, and into nitrous oxide, a powerful greenhouse gas and affect our climate. When rN and phosphorus, are in excess, they can cause multiple kinds of damage if not properly managed. They are serial polluters.

Nutrients in water, unless taken up by plants or leached into aquifers, will eventually make their way to lakes, waterways, estuaries, and near coastal waters. There, under certain conditions, they can lead to excessive algae growth and eutrophication, thus impairing the quality of aquatic habitats. They are not just transported by water to these aquatic systems. In the Chesapeake Bay, for example, one-third of the nitrogen load into estuarine waters actually comes by air transport, in its volatile phases (U.S. Environmental Protection Agency, 2015). Forests in the western regions of the United States are adversely affected by nitrogen air pollution from livestock operations upwind of the forests (National Oceanic and Atmospheric Administration, 2000). The nitrogen pollution can affect not just the leaves of the trees upon which they fall but also the general health of the vegetation by altering the pH of the soils in which the plants are growing.

There are many other ways in which altering the carbon, nitrogen, phosphorus balance affects aquatic and terrestrial habitats. Discussion of these is beyond the scope of this introduction to these natural resource management issues. We give a few references to more detailed studies of these impacts in the Recommended Resources at the end of the book.

Whereas soil erosion can often be readily observed, nutrient movement offsite cannot be observed visually. We can see fertilizer pellets being scattered on the ground when we fertilize our lawns. In time these also disappear without our seeing where the fertilizer goes.

6.2.2.1 Nutrient transport by water

As noted above nutrients present in the landscape are present in soluble forms or attached to soil particles. Water transports nutrients (and other chemical compounds) in both forms although the processes differ as does the chemical availability of the nutrients. Soil erosion transports nutrients attached to soil particles. Because we can observe the soil particles moving, it is relatively straightforward to trace how water moves these nutrients and understand their pathways. Our earlier discussion of soil erosion identified factors that contribute to soil erosion and conservation practices that can be used to reduce erosion. Actions that reduce surface runoff and the velocity of surface water runoff will reduce both erosion and the transport of the nutrients attached to the soil. Because these nutrients are bound to soil particles, they are generally less unavailable to plants or other biological activity. It is important to remember these compounds remain in the landscape and can eventually be released into the environment.

Nutrients in soluble forms are highly mobile, able to be transported in solution by surface runoff. Nutrients in these forms are more chemically active, are generally available for biological uptake, and are associated with several environmental concerns including eutrophication, alteration of sensitive ecosystems, and **hypoxic zones**. We discuss several of these in the next chapter. Excess nutrients are also associated with human health concerns, including blue baby syndrome (infant methemoglobinemia) and neurotoxins released from some algae blooms. Blue baby syndrome can occur should babies consume water with high nitrate concentrations (Christiano, 2017).

Following the transport of dissolved nutrients requires tracing water infiltration and its movement through subsurface pathways. Nutrients at the soil surface can be transported as water infiltrates into the soil. Infiltration also exposes nutrients in the soil to water transport. In both of these scenarios, subsurface water transports nutrients to waterways and aquifers.

The subsurface transport of nutrients shows that reducing surface runoff and increasing infiltration can increase the removal and transport of soil nutrients. This results in a tradeoff between erosion and nutrient movement. Conservation measures that slow water movement and increase infiltration reduce the transport of dissolved nutrients in runoff. Measures that reduce soil exposure to erosive forces such as conservation crop rotation, cover crops, protective vegetative covers, reduced tillage, and residue management will reduce soil movement and the transport of nutrients attached to the soil. However, the increased water infiltration can result in dissolved nutrients taking subsurface pathways into waterways or aquifers. The potential of nutrients to be transported in water through multiple pathways requires that we carefully follow their movement through the environment. Additionally, the mobility of nutrients raises important concerns regarding appropriate strategies for applying crop nutrients, designing production methods to retain nutrients within agricultural systems, and developing effective conservation systems to capture nutrients before they enter waterways.

6.2.2.2 Conservation to reduce excess nutrient transport

Conservation tillage and residue management systems provide protection from erosive forces in part by increasing infiltration and increasing the organic content and soil water storage capacity. These practices reduce soil erosion and nutrient transport in surface runoff and increase water storage in the organic layer.

Traditional tillage systems, on the other hand, bury crop residue and expose soils to water erosion. This results in unimpeded surface runoff. They also expose soil carbon to oxidation, reducing the organic content that binds particles together.

The benefits of conservation tillage in reducing nutrient transport in surface runoff can be offset by increased subsurface transport. Ignoring this tradeoff overstates the effectiveness of these practices in reducing nutrient delivery to waterways.

Crop rotations are adopted with profits in mind, but are also often designed to control erosion, manage disease, and enhance soil productivity. These rotations are designed to limit soil exposure to erosive forces and reduce the need for inputs, including fertilizer. Reducing fertilizer applications by including legumes that fix nitrogen and crops that require lesser amounts of nutrients can reduce the nutrients available to leave cropland. Incorporating cover crops into a rotation protects the soil. Cover crops take up and hold nutrients on site for use by the primary crop, add carbon to the soil, and promote soil health.

Nutrient management systems are effective in limiting the amount of nutrients available to be transported offsite. Well-designed systems apply the proper amount of each nutrient, accounting for all sources available to the crop, including residue in the soil, deposited from the atmosphere, and in applications of fertilizer and manure. These systems apply the proper fertilizer at the time best suited for plant growth, in the manner that supplies the nutrients to the plant with the least exposure to being transport off-field (Natural Resources Conservation Service, 2020b). Similarly, the volatilization of nitrogen applied as fertilizer can be reduced with the adoption of sound fertilizer management practices.

Grass filters and **riparian forest buffers** are conservation practices that can intercept sediment, nutrients attached to the sediment, and nutrients in runoff before they reach waterways. The effectiveness of buffers is a function of their width, supporting upland conservation practices, and the buffer vegetation. When buffers are appropriately designed and installed to handle surface runoff, they are effective in greatly reducing the sediment and nutrients reaching waterways (Kleinman *et al.*, 2018). Buffers can be also effective in reducing the nutrients flowing into the stream from subsurface water flow if the flow comes in contact with the vegetation in the root zone of the buffer. Buffers that are not sufficiently wide to handle expected runoff, are bisected by concentrated flow, or have inadequate vegetative cover may somewhat reduce the sediment and nutrients reaching the waterway, but will not be effective in protecting the waterway (Kleinman *et al.*, 2018; Wallace *et al.*, 2018). If the subsurface flow does not come in contact with the root zone, no or little plant uptake will occur and nutrients will pass under the buffer unimpeded into the waterway (Brooks & Jaynes, 2017).

Drainage systems accelerate soil drainage and move water more rapidly into waterways. The more rapid soil drainage removes water from macropores, which in turn increases infiltration during and immediately after rainfall. The increased infiltration reduces runoff and the associated erosion but increases the water transport of nutrients within the soil. Drainage systems move these nutrients directly to waterways bypassing soil processes and conservation measures such as buffers that could retain or transform the nutrients. Drainage systems contribute to higher levels of nitrogen and phosphorus being delivered to waterways fertilizing aquatic plants, and leading to eutrophication. When increased aquatic vegetation dies, the microbes that decompose the plant material lower oxygen levels in the water column. The consequence is hypoxia.

For cropland with tile drainage systems, the buffers discussed typically provide a smaller reduction in nutrient and sediment delivery. This is because with increased infiltration, there is less runoff for the buffer to intercept. On the other hand and more importantly, the drainage system bypasses the buffer to deliver soluble nutrients directly into the waterway. A recently developed conservation system, saturated buffers on tile drainage systems, increases the effectiveness of buffers in reducing nutrient delivery to waterways. A saturated buffer system spreads water along the length of the buffer using perforated pipes (Figure 6.6). The water then passes through the root zone of the buffer allowing the vegetation to take up the nutrients and denitrification to occur. Saturated buffers are not suitable for all locations because they require soils with a sufficient carbon content ($>1.5\%$), a soil horizon where the water level can be raised to pass through the root zone, and a landscape where the neighboring land will not be inundated. When properly sited, saturated buffers promote denitrification and can remove 40–60% of the nitrate in tile drainage water (Brooks & Jaynes, 2017).

Bioreactors also reduce nitrate nitrogen in drainage water (Greenan *et al.*, 2009; Woli *et al.*, 2010; Christianson *et al.*, 2012; Rosen & Christianson, 2017). The drainage water is directed through the bioreactor, an underground structure containing a source of carbon such as wood chips. As with saturated buffers, anaerobic conditions are created for efficient denitrification. Bioreactors are useful on sites where saturated buffers are not feasible because they do not require a buffer, they provide their own carbon source, they can be used in soils with a low soil carbon content, and they are not susceptible to flooding neighboring land.

Constructed and restored wetlands can increase denitrification and its efficiency (that is the proportion of N_2O released as a byproduct) with the adoption and appropriate siting of conservation practices. By creating the conditions for efficient denitrification, constructed wetlands that have well-established wetland vegetation to receive nitrate rich water can reduce nitrogen (in the form of nitrates) up to 1,570 kg for each hectare (1,400 pounds per acre) of wetland (Hyberg *et al.*, 2015). Typically, these wetlands are sited to receive agricultural drainage water. Similarly, well-designed and sited saturated buffers, restored wetlands, and bioreactors can reduce nitrate nitrogen in the water leaving farmland.

Rarely does a successful plan include only single practice. A site-specific system of production and conservation practices designed to address farm profitability, soil

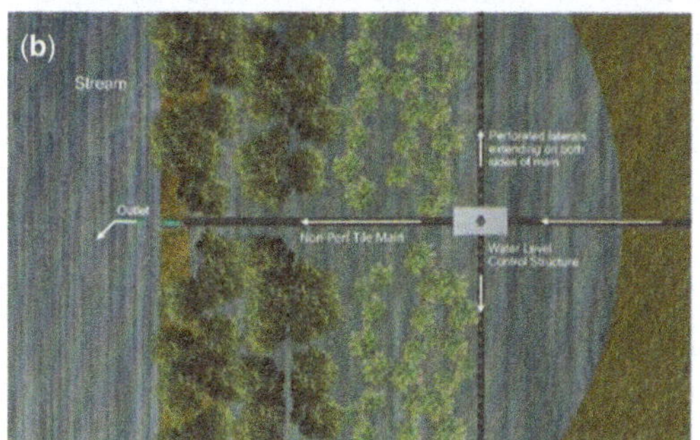

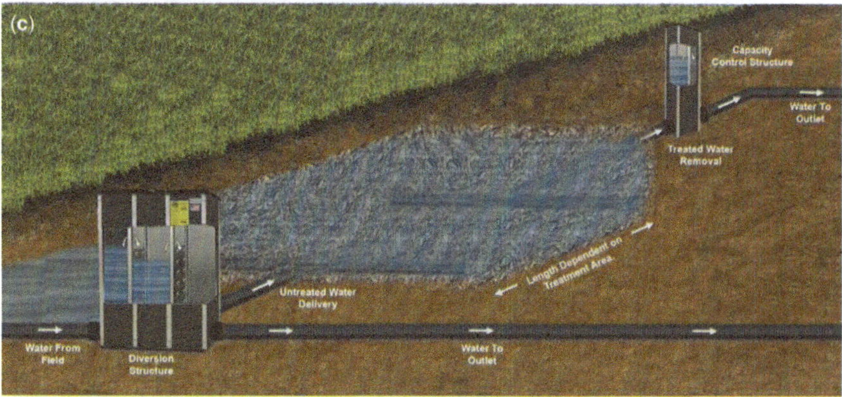

Figure 6.6 (a) Buffer bisected by a tile drain, (b) saturated buffer system, and (c) bioreactor.

erosion and health, and offsite movement of sediment and nutrients is the most effective way to balance these objectives. The guidance of a trained professional familiar with both crop production and conservation is important in developing a sound plan.

6.3 AIR QUALITY

Air quality is not spared by imbalances of carbon, nitrogen, and phosphorus. Imbalances manifest themselves in three general ways: increased dust in the air, increased particulate levels in the air that can damage the lungs, and production of noxious compounds that affect smell or human welfare. As explained in the discussion of wind erosion, soils that lack sufficient organic matter to bind soils into larger aggregate particles are vulnerable to water and wind erosion. In the United States 29.8 million hectares (73.6 million acres) of land have been identified as vulnerable to wind erosion (U.S. Agricultural Research Service, 2020).

Perhaps the best-known incidence of wind erosion is the Dust Bowl from the 1930s in the American Midwest. The 'black blizzards' made life intolerable for millions of Americans, spurred a mass migration, and destroyed millions of acres of wildlife habitat and cropland (Figure 6.7).

The second way in which a carbon–nitrogen–phosphorus imbalance affects air quality is through the production of air particulates. Fertilizer applications and manure emissions originating from intensive livestock operations emit volatile forms of rN. Ammonia and NOx (which results from the burning of **biomass**)

Figure 6.7 Dust Bowl in USA Great Plains in the 1930s. *Source*: Wikicommons (USDA Soil Conservation Service).

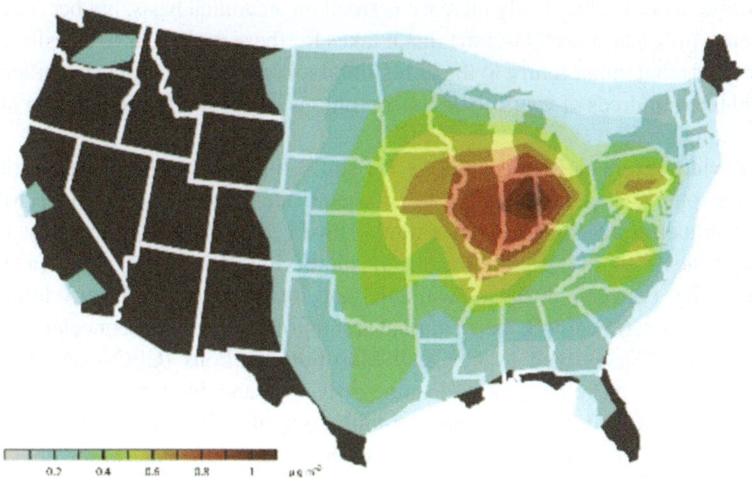

Figure 6.8 Areas with elevated ammonia concentrations *Source*: Wikicommons (NASA).

create **inorganic** aerosols, generally ammonia-containing compounds. The small particulates (referred to as particulate matter (PM) 2.5 whereby the units are microns) are a major health concern in the United States, Europe, and the rest of the world (Bauer *et al.*, 2016). Figure 6.8 shows the areas of the United States with elevated ammonia concentrations in the air. These occur largely in the agricultural Midwest and MidAtlantic states where much of the pork and poultry are produced.

We all have passed by intensive livestock operations, such as a large hog, chicken, or dairy farm. The odor we smelled was most likely ammonia and hydrogen sulfide, another breakdown product of protein from dead organisms and waste. The odors emanate from the operations themselves where many animals are produced in a confined space generating large amounts of animal waste and from the fields where the manure has been applied. Not only is the odor unpleasant, but elevated concentrations of it are also a threat to human health. As a final insult, the odors can reduce the value of property downwind imposing an additional economic cost.

6.3.1 Climate impacts

Disturbing the balance between carbon, nitrogen, and phosphorus pools can have impacts that are both lagged and long-lasting. Increased atmospheric emissions of the greenhouse gas components of the cycles lead to their steadily increasing concentrations in the atmosphere. In other words, there are shifts in mass of carbon and nitrogen from the soil or other long-term storage pools to the atmospheric pool. This steady increase is small on an annual basis, but because their residence time (the amount of time that it takes for these molecules to transform into compounds that either return to the terrestrial pool or shift to long-term storage) can extend into hundreds of years, their concentrations can grow. The impact can last for a very long time.

As evident from earlier discussions, the shifts in mass from long-term storage to the atmospheric pool are primarily the consequence of mining, fossil fuel drilling, and chemical conversion (N_2 to nitrate). However, a portion of the shift to the atmospheric pool is also the result of modifying the natural landscape that reduces the conversion of bioavailable and reactive material into long-term storage. Figure 6.9 shows the historical trends of atmospheric concentrations of CO_2, N_2O, CH_4, as well as CFCs. [Chlorofluorocarbons (CFCs) are synthetic chemicals used as refrigerants, cleaning solvents, and blowing agents.] Global efforts have succeeded in reducing the emissions of CFCs in recent years. As is obvious from the other three graphs, we have had little success with carbon and nitrogen.

Anthropogenic carbon enters the atmospheric pool as carbon dioxide where it acts as a greenhouse gas, and as methane, another greenhouse gas. Anthropogenic nitrogen enters the atmosphere primarily as nitrous oxide but also as ammonia. Ammonia, with a very short lifespan in the atmosphere, is not considered a greenhouse gas. The carbon dioxide comes primarily from combustion of carbon-containing products, such as fossil fuels, or from the breakdown of the

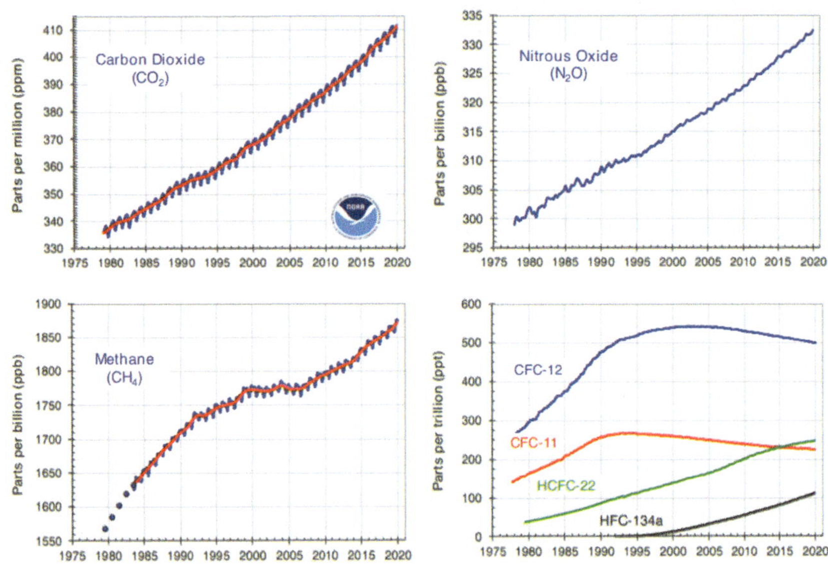

Figure 6.9 NOAA Global Monitoring Laboratory (2020).

organic component of soils. The additional methane comes largely from the fossil fuel industry and leakage from natural gas transport. It also comes from livestock operations and from fields flooded for rice production. Incomplete conversion of nitrate or ammonia to N_2 through microbial action and volatilization of fertilizer can lead to the production of nitrous oxide, a powerful greenhouse gas. This phenomenon occurs in both soils and in nutrient-enriched wetlands, estuaries, lakes, ponds, and other similar waterbodies where water flow is impeded and rN can concentrate. What percent of total rN applied as fertilizer eventually converts to N_2O is unclear, but is often estimated to be around 4% (Xiao *et al.*, 2019). Some thirty million tons of N_2O enter the atmosphere each year, of which about 36% is the direct result of human activity (U.S. Environmental Protection Agency, 2010). This estimate is in dispute because a growing amount in recent decades derives from melting permafrost soils of the arctic boreal regions. Some argue that any nitrous oxide emissions resulting from human-induced warming of these regions, i.e. climate change, should also be attributed to human activity (McDermott-Murphy 2019).

Because rN is a serial offender, nitrous oxide derives not just from a number of pools, but also from a number of other impacts. Oceanic and estuarine dead zones are also sources of nitrous oxide emissions, as are reservoirs behind dams. Ironically, a dam which is intended to serve to reduce the use of fossil fuels and hence the generation of greenhouse gas emissions instead can become a major source of greenhouse gas emissions, that is, nitrous oxide, because of the concentration of nitrates and ammonia in the slow-moving waters behind the dam. This should remind the reader that even what seems to be the best solution to a problem needs to be thoroughly examined to avoid unintended consequences.

Phosphorus exists as a liquid or solid at normal environmental temperatures. Hence, of the three ingredients of life, only phosphorus does not also have a gaseous form and thus contribute to the atmospheric pool.

6.4 SUMMARY

We humans have changed our world to exploit the benefits of its natural resources. We have changed the flow the streams and rivers and our technological innovations have altered carbon–nitrogen–phosphorus cycling. The consequence has been that the mass in long-term storage pools has become available for human use in terrestrial pools. What was once scarce has become readily bioavailable.

We accrued the benefits upfront. The costs have been often delayed or unexpected. Neglecting to consider where water goes and the consequences of expanding the bioavailable pools of carbon, nitrogen, and phosphorus have led to sticky problems. The result of this neglect is increasing damages from flooding, polluted air and water, ocean dead zones, the slow and steady rise in atmospheric greenhouse gases, and falling pH levels of the oceans. Dealing with these issues poses significant, hard-to-resolve challenges.

Our intent in describing these impacts is not to depress the reader into inaction. Rather it is to make you aware of the potential consequences of ignoring the principles that everything goes somewhere, gravity works, and natural systems will follow their rules to reestablish balance. Taking account of these principles can help you make better decisions in examining measures for action as well as evaluating the decisions of others. We all want to avoid avoidable mistakes.

It is difficult if not impossible for any one of us alone to solve the regional environmental and natural resource problems, let alone the world's problems, that confront us. But we can make local problems that we see around us better and avoid contributing to future problems in our immediate environment. These tools should help. In any case, you should have a better understanding of the consequences of ignoring nature and its laws.

In the next chapter we present examples of how not thinking out the implications of actions have led to undesired consequences. We help the reader identify what could have been done differently to have produced better outcomes.

Chapter 7

Putting it all together: case studies

There was an old lady who swallowed a fly
I don't know why she swallowed a fly – Perhaps she'll
die! There was an old lady who swallowed a spider;
That wriggled and jiggled and tickled inside her!
She swallowed the spider to catch the fly.....

—Rose Bonne, 1961

The old lady of the nursery rhyme mistakenly swallows a fly and then eats a series of ever larger animals (spider, bird, cat, dog…) to 'solve' the problem of having eaten the fly. Sometimes environmental management can seem to follow the same strategy. The problem to be addressed often results from actions taken in the past. The potential solutions to that problem need to be carefully analyzed to avoid creating a new problem that needs to be solved. To be fair, greater understanding of how natural systems function has identified some of the problems generated by previous solutions. But in a number of cases these might have been avoided had our basic principles been were followed – gravity works, everything goes somewhere, cycles balance. [Disturb at your own risk!]

So far, we have looked at and brushed up a few basic concepts you already knew, discussed cycles important for understanding environmental concerns, examined natural and human change within our world and the impacts of that change, and introduced a few tools that can be used to help account for human activity. We now present case studies whereby you can begin to test where well-intended actions to solve problems have gone awry. In other words, the action resulted in

© IWA Publishing 2021. A Guide to Understanding the Fundamental Principles of Environmental Management. It Ain't Magic: Everything Goes Somewhere
Authors: Andy Manale and Skip Hyberg
doi: 10.2166/9781789060997_0125

undesired environmental outcomes or the decline in the quality of the natural resource. These case studies illustrate what happens when the basic concepts which we introduced to you at the beginning of this book are neglected. In every case, human modifications caused disruption of the natural system, the consequences of which could have been anticipated. We encourage you to consider what individual and collective actions (public policies) can be taken to correct for these mistakes. We touch upon this subject in the next chapter.

First, we want to introduce you to the concept, important for the management of natural resources, of what is referred to in the scientific literature as 'the Commons.' The idea relates to the science to which we have introduced you and to the circumstances in which we, as one of many sharing the planet with our fellow humans, plants, and animals, live.

7.1 THE COMMONS

The Commons is a resource that we share and upon which we depend for sustenance, recreation, health, and more. It is a resource that is shared because its use cannot be appropriated by any single individual or entity. When individual users of the shared resource, acting in their own self-interest, deplete or cause a diminution of the quality of the resource, all users suffer, we have what is known as the 'Tragedy of the Commons,' a term coined by Garrett Hardin (1868). Common resources are not controlled by any party, so no entity has a vested economic interest in investing capital and labor into protecting and managing the resource because the returns from their investment will be consumed by others. Most environmental and natural resource problems, including those sticky ones discussed in this book, are often referred to as 'Commons' problems.

For the purpose of explanation, the Commons is often presented as a place, such as a forest, grasslands, or fishery, that is used by diverse woodsmen, shepherds, or fishermen. Overexploitation by one entity, or more often many entities, affects the health and productivity of the resource for all – hence the tragedy of misuse, resulting from pursuit of self-interest. More recent thinking extends this concept to resources such as air, which we share globally, or water bodies that many individuals or communities use. These resources are something that we can visualize, touch, or measure.

We generally do not think of a physical or chemical state as a common resource, but all life depends on and benefits from maintaining the balancing among carbon, nitrogen, and phosphorus pools. Maintaining the balance between the carbon in the atmospheric pool (as the gas carbon dioxide) and the terrestrial pool of chemically complex carbon-containing compounds, whether it is held as a mineral, sequestered in soils, or stored deep in the earth as deposits of fossil fuels, stabilizes multiple chemical cycles. Similarly, a balanced pool of reactive nitrogen benefits higher organisms, including humans. Excess availability of rN would play to the advantage of single-celled organisms. A similar relationship exists with phosphorus pools. And the availability and abundance of water in its non-saline (salty or briny) fluid form determines the productivity of our terrestrial world. Our species depends upon the balancing of these critical elements of the biosystem. This is balancing that occurs within chemical species and across chemical species. Hence, like the shared grassland, the balancing of these elements of life is in our 'common' interest. Situations where individual self-interest lead to the Tragedies of the Commons include exploitation of the carbon pool stored as fossil fuels and its conversion into carbon dioxide, a greenhouse gas; creating new reactive nitrogen from inert reserves in the atmosphere and releasing it in its myriad forms across air and water pools; mining phosphorus in its form as phosphate from mineral reserves and making it bioavailable where it would naturally not be; and altering the natural flow of water to the detriment of ecosystems. Each of these examples moves a mass of an element (or in the case of water, a molecule) from one pool to another, creating an imbalance in one chemical species and often disturbing the balance of the other chemical elements of life.

The environmental and natural resource problems discussed in this book are associated with modern technologies that have made what was once scarce plentiful. The opening of our modern Pandora's Box is the actions of individuals working in their own self-interest producing outcomes that have had tremendous public benefit, but detrimental side effects. These negative consequences are shared by all, yet no one individual or collection of individuals wishes to bear the burden of the cost of **mitigation** or forbearance necessary to restore balance. This is a critical barrier to solving the problem of the 'the Commons' and thus it impedes resolution of the tragedy.

7.2 CASE STUDIES

To this point we have been pushing you to build upon what you already know and the skills you have learned at an early age. You should be able to use your knowledge to track the consequences of changes on the land. We encourage you to ask questions when observations deviate from the predicted outcomes. These consequences we have discussed include the effects from changing the mass of carbon, reactive nitrogen, and bioavailable phosphorus (and their forms) in one pool and shifting it to another, the impacts of modifying how water moves across the land, and the effects of these changes on downstream systems. We have been preparing you to recognize the strengths and weaknesses of proposed management options and identify those proposals that are either incomplete or poorly conceived and need further examination to avoid unintended consequences before proceeding to action.

In the following case studies you can begin to review cases where collective action has gone awry resulting in undesired environmental results diminishing natural resource quality. We hope that these examples will help prepare you to be able to anticipate problems that can arise and learn to how to avoid similar outcomes. We also hope that after working through these cases you will feel confident that you already have a solid base to build on as you continue your education.

7.2.1 Some background

In our discussion of 'the Commons,' we explained how many small individual actions can sum to a large impact that degrades the quality of natural resources. What follows is an actual illustration of the concept. This foreword will help illuminate the subsequent case studies.

In 1862 in the United States, the Homestead and the Pacific Railroad Acts became law. The Homestead Act permitted any adult to claim 65 hectares (160 acres) of land by improving the tract by growing a crop and building a home. This act along with several other later Homestead Acts transferred over 65 million hectares (160 million acres) of federally owned land to private landowners. The Pacific Railroad Act and subsequent legislation were adopted to construct a continuous railroad line between the Mississippi River and the Pacific Ocean. These acts conveyed over 70 million hectares (175 million acres) of public land to railroads. Much of this land was sold to settlers to fund the construction. The settlers, each acting independently, converted grasslands, wetlands, rangelands, and forests into cropland. Each farm by itself had a minimal impact on the environment. In the aggregate, however, these conversions dramatically altered the ecosystem of the Midwestern United States. The extent of the change is reflected by the fact that less than 1% of the tall-grass, 30% of the mixed-grass, and 75% of the short-grass prairie remain (Knopf & Sampson, 1996, National Park Service, 2020a, b).

The consequences of this conversion are numerous. It altered the hydrology by changing the soil infiltration capacity, the characteristics involving evapotranspiration, and in many cases how much and how fast water drains. Cultivation (tilling) exposed the soils to water and wind erosion. This in turn increased sediment delivery along with attached nutrients to streams, rivers, and lakes. It also exposed soil carbon to air, leading to its oxidation and the depletion of soil organic matter. Hence, carbon that once was confined to the longer-term storage pool of soils was released as CO_2 into the atmosphere (the atmospheric pool). The draining of wetlands decreased the capacity of the land both to retain water after storms and to convert rN back to its inert form. All of these changes to the hydrology increased peak river and stream flows. Nutrients that would have otherwise remained in the soil profile were discharged into waterways where even now they make their way down the Mississippi River to the Gulf of Mexico adding to one of the largest hypoxic zones in the world and thwarting contemporary efforts at mitigation. And over time as fertilizer and other agricultural chemicals came into use, the hundreds of millions of acres of cropland became a nonpoint source of pollution for the United States waterways.

Each of these changes has wide-ranging impacts on natural resources and the environment. We will subsequently explore a few in the case studies.

One increasingly important consequence is the decline of habitat quality for native flora and fauna. We note this impact, but because the discussion falls outside of our examination of environmental management, we provide references for further inquiry. With the loss of the flora came the disappearance or major population declines of many of the native animal species. We will not be studying these impacts within this text, but will note them in several case studies.

7.2.2 Case study 1 – the Great Flood of 1993: gravity matters and water goes somewhere

The Great Flood of 1993 inundated millions of acres of agricultural land in the midwestern region of the United States, causing $26.6 billion in damage to urban areas. It was the costliest flood in the history of the United States history up to that year. Some 78,000 square kilometers (30,000 square miles) were flooded, 830,000 square kilometers (320,000 square miles) were affected, and 50.000 homes were destroyed. Nature set the table by providing storms during the late spring and summer that pounded the Upper Midwest with seemingly unceasing rainfall. Rainfall was 400–750% above normal in some areas of the northern-central plains. Parts of east-central Iowa received about 48 inches (120 cm) of rain between early spring and late summer (Interagency Floodplain Management Review Committee, 1994; Larson, 1996; National Space and Atmospheric Administration, Earth Observatory, 2005).

For an event this large extensive flooding was inevitable, significant damages would have occurred no matter what, but human actions greatly increased the amount of damage. Perhaps the most consequential was the placing of homes, high-valued buildings, businesses, and infrastructure on land that was within the floodplain. A flood plain clearly outlined by the 1927 flood. The management strategy of constructing levees to protect these manmade structures created a false sense of security that encouraged additional construction in the floodplain. All of this assured catastrophic damage when a major flood eventually did happen.

Drainage systems installed by farmers quickly removed the waters from their fields. This water had to go somewhere. Satellite images of the region taken over a period of weeks after the downpours show where it went. The pictures show inundated cropland over a very wide area during the early days of the flood. Subsequent images in the days that followed tracked the floodwaters to the confines of the tributaries and the mainstem river floodplains. The water had to go somewhere. It flowed into already swollen floodplains and downstream areas, exacerbating the flooding. Remember that although only a portion of the basin experienced rainfall amounts exceeding a hundred-year event, all the rain that fell on land within the greater watershed eventually made its way into the mainstem of the Mississippi River.

Levees increased the damage. These levees (particularly agricultural levees) that had been constructed along tributary rivers and the mainstem far from cities prevented the waters from going onto floodplains. The water had to go somewhere, so it continued downstream where it was squeezed higher and higher by the levee systems. Eventually more water was forced between the levees than they could handle and it rose to the level where it overtopped and, in some cases, breached them. The system designed to mitigate major damage to public health and welfare was overwhelmed. So even though decades before, the 1927 flood demonstrated the futility containing flood waters within a levee system during a major flood event, the lessons had not been learned. Constraining floodwaters resulted in waters rising ever higher. When the area between the levees filled, the remaining water had to go somewhere. High volumes of floodwater and gravity returned the water to where it wanted to go. The result? Towns and cities in the floodplain flooded.

An analysis of the flood found that a landscape modified by cropland drainage systems had altered the hydrology of the Mississippi River basin (MRB) (Interagency Floodplain Management Review Committee, 1994). Draining wetlands and converting natural vegetative covers to agricultural crops greatly reduced the capacity of the upland portions of the basin to slow, store, and gradually release water. These modifications increased an already large volume of water, which would normally have traveled more slowly from upland to downstream areas, raising the water level along the mainstem, reinforcing the effect of the levees. The result of the more rapid downstream movement of the flood waters and the constriction of the water within the levee system had calamitous consequences.

What do our principles say to do? Gravity and water will have their way. One cannot predict at any one time the magnitude of rain events, but one can estimate their probabilities and plan accordingly. Common sense also plays an important role. Given that a major flood had taken place 67 years before, observing its impacts would provide a guide for future events. Environmental managers would be wise to plan for providing a system to estimate the volumes of water that will result from a major precipitation event. The system should have the capacity to handle that flow volume safely. Levees can have a role in protecting population centers and high-valued infrastructure, but over reliance on levees to protect all activity within floodplains can overload the capacity of a river system to handle major floods. In the design of this system, managers need to take into account the landscape modifications that have taken place that reduce storage volume. To do otherwise is to tempt fate and the probability of highly costly damages.

7.2.2.1 Take home lesson

The human actions that made the Mississippi Flood of 1993 as severe as it was were cumulative and over a century in the making. Draining large areas of wetland and accelerating drainage from cropland both contributed to raising peak flows by delivering water more quickly into the Mississippi River and its tributaries. Flooding is a function of volume of water and timing; the latter is determined in part by how fast the water moves downstream. Soil and land attributes affect the timing. Volume is affected not just by the intensity of the rainfall events but also vegetation and hence the rate and volume of evapotranspiration. Levees can protect areas from flooding during small and moderate events, but can lead to disasters during major floods. Separating the floodplain from the river with levees increased the water velocity and further increased peak flows. The higher velocity and water level increased the force the river was able to exert against levees and increased the likelihood the levees would either be overtopped or breached. The modification of the vegetative cover served to reduce the water holding capacity of the soil which in turn increased surface runoff and the peak flows. Planning for the purpose of mitigating flood damages should therefore address what is grown upstream and when, land attributes that can change the movement downstream (including surface and tile drainage systems), and access or connectivity of rivers to their floodplains. Not situating vulnerable buildings and infrastructure within the floodplain can also reduce damages. Environmental managers will spend much time and effort working to ameliorate changes within landscape, redesigning drainage systems, restoring and constructing wetlands, and reconsidering floodplain management for the Mississippi River and its tributaries.

7.2.3 Case study 2 – watershed management: Crabtree Creek Raleigh, NC

Crabtree Creek watershed encompasses some 142 square miles in the state of North Carolina, USA. The upper third of the watershed has broad expanses of gently rolling upland with wide flat floodplains; the lower two-thirds of the area are gently rolling to rolling upland, with steep breaks adjacent to streams and irregular width floodplains. In 1964, cropland represented 14% of the land area, 14% was urban, woodlands covered 60%, and 3% were devoted to pasture and hay. The headwaters of the watershed experienced flooding two or three times a year with at least one flood occurring during the growing season (Neuse River Soil and Water Conservation District, 1964).

Flooding in the Crabtree River valley is a natural occurrence reflecting variability in rainfall patterns, with the additional volumes of water spilling out onto floodplains and into wetlands. Flood damages, on the other hand, result from humans putting structures where floodwaters normally would go or conducting activities in areas inconsistent with the likelihood of inundation or high waters. Growing crops that cannot tolerate even short periods of wetness is an example.

What has happened in the watershed since 1964 reflects ever greater use of the land in ways that put human health and welfare at risk from flooding. Population in Raleigh has grown by over 290% (World Population Review, 2020). Land area available for temporary water storage has commensurately declined. Floods causing significant damage have occurred during a major precipitation event in June 1973, Hurricanes Fran (September 1996), Dennis, Floyd, and Irene (September–October 1999), Matthew (October 2016), Florence (September 2018), and tropical storm Alberto (June 2006) (Bales *et al.*, 2000). However, it does not take a major storm for Crabtree Creek to flood. In fact, since 1970 Crabtree Creek has flooded frequently from less-intense rain events, with substantial economic impacts. The severity of the damages has been exacerbated by development within the floodplain and the watershed.

Population growth and the associated increase in infrastructure contributed to these damages, but two basic mistakes greatly exacerbated the damages: inaccurate population projections and development within the floodplain. The first mistake was with projections for growth in the watershed. Actual population growth far exceeded the projections used by the planning assessments developed in the 1960s, leading to the risk of flood impacts being underestimated. The underestimated population is puzzling. The Environmental Impact Statement (EIS) for the Crabtree Valley estimated population growth to be 114% over 50 years, a rate that was less than half the actual growth rate observed between 1964 and 1973 (U.S. Environmental Protection Agency, 1976; Neuse River Soil and Water Conservation District, 1964; Tuttle, 2013; U.S. Census Bureau, 2020). Data from the U.S. Census Bureau (2020) and the Federal Reserve Bank of St Louis (2020) indicate the population for Raleigh would grow 292% and Wake County grew 385% between 1970 and 2020 (estimated). These estimates are far more in keeping with the growth observed between 1950 and 1970 than what were used by the EIS. Better predictions on population growth were available when the EIS was prepared.

A consequence of the erroneously low prediction of population growth was that the planning for the watershed underestimated the amount of land converted from woodland and cropland to buildings, roads, and other impervious landcovers. As we have discussed, everything goes somewhere. If the water cannot infiltrate it will become runoff. And because the impervious surfaces provide more runoff and less obstructions, more water runs off faster. Additional impervious surfaces in a watershed with a documented history of flooding exacerbated the frequency and level of the flooding.

Because the amount of impervious landcover was underestimated, the peak flow for rain events was underestimated. This led to underestimating the risk of flooding to businesses located within the floodplain and correspondingly the extent of damages from that flooding. The repeated flood damage to Crabtree Valley Mall, built in 1972, and adjacent businesses including several car dealerships, illustrates the consequences of this underestimation. Commercial insurance available to the Crabtree Valley Mall has been sufficient to cover flood damage for only two stores (Tuttle, 2013).

Policies and ordinances dealing with impervious covers and construction within the floodplain might have avoided the development within the Crabtree Creek floodplain which would have prevented much of the damage associated with Crabtree Creek flooding. The City of Raleigh is working to reduce and mitigate damage from flood events by restricting development within the core of the floodplain to prevent increased damage to other watershed residents and to permit activities such as agriculture, forestry, lawns, parking areas, golf courses, parks, and other recreational uses. It has also adopted policies to increase potential temporary water storage by removing several structures within the floodplain and replacing them with parkland that can serve to hold stormwater during rain events (Raleigh News and Observer, 2006; Tuttle, 2013).

7.2.3.1 Take home message

Gravity and storage capacity matters. Water has to go somewhere. Any development that decreases temporary water holding capacity in a watershed will increase the risk of damages from more frequent and higher water levels. Increasing impervious land cover will reduce the water storage capacity of soils within the watershed and increase the volume and speed of runoff. Placing structures in places where gravity dictates where water should go, converts a flood event into an event involving flood damages. Much of the damage from flooding that now occurs is due to placing structures within a known flood zone, reduced soil water holding capacity, and inadequate planning based on questionable population projections. Damages could have been avoided by accounting for additional and unpredicted growth in the watershed and taking appropriate actions, such as those that have been adopted recently in Raleigh. A final message from this example is this: examine and challenge assumptions. A more careful examination of projected population growth might have resulted in less construction within the floodplain.

7.2.4 Case study 3 – the Des Moines River watershed: a more integrated examination

The Des Moines River watershed in the state of Iowa and the heart of the American Corn Belt has highly productive agricultural soils, a beneficial climate, and extensive tile drainage systems. Iowa State University, located in this watershed, has extensively monitored and studied the impacts of the conversion and use of these soils in intensive agriculture. These studies provide a more complete understanding of how vegetative cover, infrastructure, hydrological function, nutrient cycling, and water quality interrelate. They help make clear how these changes to the natural landscape affect both water flow and impact water quality.

Iowa is a leader in corn and soybean production where farmers use a substantial amount of fertilizer to achieve high yields. It also has a highly drained landscape. In fact, Zulauf and Brown (2019) noted that Iowa is the state with the highest percentage of cropland with tile drainage (53%) (see also, National Agricultural

Statistics Service, 2019a). The draining of wetlands, the associated tile draining systems, and the development on the floodplain have each played a role in increased flooding in the city of Des Moines, with major flooding occurring in 1993. The combination of a highly cropped, fertilized, and drained landscape has resulted in impaired water quality, with drinking water for the major city in the watershed, Des Moines, regularly exceeding the drinking water standard for nitrates (Des Moines Water Works, 2020). Des Moines Water Works had to install an expensive water purification system that cost several million dollars in order to meet these standards (Des Moines Water Works, 2020). In 2015 the equipment was in operation 177 days costing $1.5 million (Des Moines Water Works, 2016).

Calls to reduce the loadings of nitrates into the Des Moines and Raccoon Rivers initially focused on decreasing nitrogen fertilizer application rates within the watershed. Though seemingly the feasible approach, simply focusing on fertilizer use failed to account for all sources of nitrates. An examination of nitrogen fertilizer use within the Raccoon River watershed between 1989 and 2003 showed little change in the amount applied. Further the increased fertilizer uptake by crops indicated increased nitrogen use efficiency raised doubts that increased fertilizer applications were the source of the additional nitrate (Hatfield *et al.*, 1999). Manure applications were considered by some as the source of the nitrate, but an examination of cattle and hog production over the period in question revealed the number of animals had declined by 63% and 20%, indicating the nitrogen available for application had decreased by 25% (Hatfield *et al.*, 1999). So why was the nitrate concentration in the river continuing to rise while the total nitrogen application decreased? Pause for a moment and think about the scientific principles and concepts that we have been examining. What do we know and what has not been examined?

We know that mass is conserved, neither disappearing nor being created. And we know water flows downhill from the uplands to the rivers where it infiltrates soils downward until it reaches the water table. Finally, remember different plants have different transpiration rates. In an examination of how agricultural practices affects carbon, nitrogen, and phosphorus cycling, Hatfield *et al.* (2009) found that crop rotations (crops sequentially grown on the same plot of land) had been steadily changing over time. As the amount of land in small grains (e.g., wheat and oats) and grasses (e.g. hay) declined, nitrate concentrations in water systems increased. This relationship between the crop rotations and nitrate concentrations can be understood by examining the timing of fertilizer applications, precipitation patterns, and comparing water use patterns and movement associated with small grains and hay with those of the corn and soybeans that replaced them.

In Iowa, April, May, and June have the three of the highest average monthly precipitation rates (rssWeather.com,2020) and are the months when spring fertilizer is typically applied. Now consider water use by these crops. Hay and small grains are actively transpiring between April and mid-June reducing the water movement to base flow. During the same period corn and soybeans have limited water use, leaving more water to move downward through the soil profile contributing base flow. Replacing small grains and hay with corn and soybeans, allows nitrates from recently applied fertilizer to be flushed out of the soil with the water and transported with the base flow to the river. The timing of fertilizer applications, and precipitation and temperature patterns contribute more to explaining nitrate concentrations in the river than changes in fertilizer use (see also Jayasinghe *et al.*, 2012).

7.2.4.1 *Take home message*

The Des Moines Watershed containing the Raccoon River Basin demonstrates the interconnectedness of the landscape, vegetative cover, water movement, and nutrient cycling, in other words how changing carbon, nitrogen, and phosphorus balances can produce undesired results. The substantial modifications of the Iowa landscape – the loss of wetlands and their water storage function, the conversion of native vegetation and wetlands to cropland, fertilizer and manure applications, and changing crop rotations – have altered the ecosystem services provided. Water pathways through the landscape (and transpired into the atmosphere) have been altered. The reduced ability of the landscape to hold water has intensified the watersheds' response to major precipitation events resulting in greater flood damage. The alterations of water movement combined with nitrogen applications on cropland have resulted in elevated nitrate concentrations in the rivers. The changing crop rotations altered plant water uptake during the period when fertilizers were applied, leaving more water to percolate through the soil and transport rN to the base flow which then carried it into the waterway. This case study shows that the interrelationships between changing land use, water movement, crop production, nutrient availability, and water quality are complex and need to be carefully examined before assigning relationships between cause and effect.

Although all of the above variables discussed play a role in the increased nitrate concentrations, this case study was presented to stress the use of a tool as well as our core principles. Many in the Des Moines River watershed jumped to what was a feasible but incorrect conclusion that increased fertilizer or manure applications were driving the increasing nitrate concentrations in the Raccoon River. By observing and measuring the amount of nitrogen applied to crops over time, this hypothesis was rejected. When an analysis taking into account water availability and movement was completed, the role of changing crop rotations in increasing the water available to transport nitrates through the soil column was documented. The lesson here is not only do we need to apply scientific principles but we also need to test our hypotheses by observing and measuring outcomes. If we reject our hypothesis, we need to reexamine the system on the basis of fundamental science.

7.2.5 Case study 4 – eutrophication in western Lake Erie

Sometimes doing seemingly all the right things still leads to unexpected trouble (and an undesirable environmental outcome) when a key principle is neglected – in this case, the conservation of mass. In this example, farmers along the western half of Lake Erie Ohio (USA) adopted conservation practices to reduce soil erosion, following recommendations from farm advisors (Jarvie *et al.*, 2017). Lake Erie had for years suffered from diminished water quality from sediment eroded from nearby farms. The increased turbidity caused by the sediment reduced the aquatic vegetation and the fish that depend upon it, diminished the recreational value, and lowered drinking water quality.

Farmers in the region adopted a large number and variety of conservation practices to help protect the water quality of Lake Erie. They used conservation tillage practices to slow runoff and to reduce erosion and sediment transport. With reduced tillage, water stays on the land longer where it can infiltrate and saturate the soil profile. On the other hand, the retention of more water on the land can reduce crop yields, leading farmers to install tile drainage. Farmers also reduced total fertilizer use and adopted cover crops to reduce nutrient runoff. Cover crops provide the additional benefit of protecting the soil outside the primary growing season and generate more organic matter that can enhance soil health.

Water quality improved from the early 1980s into early years of the new millennium. Then water quality problems began to recur. This time sediment was not the predominant water quality problem – it was algal blooms and eutrophication. These algae, actually cyanobacteria, can produce a toxin that can sicken people and kill pets (Smith, 2020).

How could practices that at first succeed, over time, fail? Achieving high yields for the crops grown on these soils requires the proper mix of bioavailable phosphorus and reactive nitrogen. To meet the nutritional needs of crops, farmers applied fertilizer. However, the no-till systems that many producers adopted resulted in fertilizer being applied near the soil surface instead of being incorporated throughout the root zone. Over time this can lead to the accumulation of phosphorus at the soil surface and the saturation of the soil particles with phosphorus. This condition leaves more phosphorus in a soluble form that can be readily transported by water.

By adopting practices that increased infiltration, soil organic content, soil water holding capacity, along with the planting of cover crops that retain phosphorus, in both solid and soluble forms (as well as reactive nitrogen) at the soil surface, farmers inadvertently increased the propensity for the transport of soluble phosphorus (and rN) through the expanded drainage systems. The increased drainage moved the water and the soluble nutrients into streams and rivers and ultimately Lake Erie. The practices that had kept sediment and the phosphorus associated with the sediment from leaving cropland soils ironically led to high accumulations of phosphorus in soils and ultimately to the transport of soluble phosphorus off-site, contaminating water bodies.

In Lake Erie waters, the combination of bioavailable phosphorus and excess rN provides the essential chemical elements for the production of life. In this case, this life is a microbial organism whose propagation in shared resources is threatening to human health and welfare.

7.2.5.1 Take home message

Even seemingly slight changes to how soils cycle carbon, nitrogen, and phosphorus can have significant consequences to associated natural resources. Successful management has to take into account all forms of carbon, nitrogen, and phosphorus that can be transported offsite. In this case, the focus had been on phosphorus associated with sediment. The practices adopted to mitigate soil erosion and the sediment containing phosphorus that was transported offsite inadvertently led to the buildup of soluble phosphorus which could be transported by water. This was taking place at the same time practices were being adopted that increased both the water holding capacity and drainage of the soils, resulting in soluble phosphorus being transported through the drainage system directly into waterways. Total phosphorus is conserved. If at a site less phosphorus is removed through the removal of residue or from loss of phosphorus through soil erosion, more will be retained until no more can be bound to soil sediment particles and it becomes solubilized. Asking the question where a nutrient (in this case phosphorus) is going (i.e., its shorter- and longer-term fate), can help lead you to identify possible problems with a strategy for longer-term mitigation.

7.2.6 Case study 5 – Erath county, Texas: dairy country

Economic considerations can lead to dire environmental and welfare outcomes if our scientific principles – gravity, the conservation of mass with the implication that water has to go somewhere, and the balancing of carbon, nitrogen, and phosphorus cycles – are ignored in decisions regarding the intensity of production. Erath County in Texas (USA) serves as an example of what happens when excess nutrients lead to the diminution of natural resource values.

Since the 1980s the dairy industry has been booming in Erath County. The county consists of two watersheds, the Bosque and the Paluxy watersheds, which are tributaries of the Brazos River. The Brazos is the main source of drinking water for the city of Waco. It covers some 1090 square miles or 2,800 square kilometers of relatively gentle sloping grazing lands within an easy drive west of Dallas and northwest of Houston.

The dairy industry grew rapidly in the 1990s, adding a new dairy every month. These were not the mom and pop diaries seen in children's picture books, but industrial farming operations (of 1,500–2,000 milking cows). A driver for this shift was the migration of dairy operations from the Netherlands to the United States and western states, a consequence of regulatory pressures to reduce water pollution in Europe and domestically (Jones *et al.*, 1993). The situation that led to a concentration of dairy operations in Erath County is analogous to the arcade game of Whack-A-Mole whereby one whacks one mole popping up from a hole with a hammer and the mole reappears from another hole (Jones *et al.*, 1993).

To feed the approximately 110,000 dairy cows maintained at these facilities between 1997 and 2002 required large amounts of feed (National Agricultural Statistics Service, 2019a). A dairy cow needs approximately 23–25 kilograms (50–55 pounds) of **dry matter** (DM) a day which includes 11 kilograms (24 pounds) of grain. The grain typically includes meal from corn and soybeans (Fischer & Hutjens, 2019). Nearly all the grain was imported from outside the watershed since there is relatively little corn and soybean production in Erath County. As we have discussed, importing corn and soybeans transports a substantial amount of nitrogen and phosphorus into this small watershed.

We estimate that during this period all dairy cows in Erath County produced manure containing nearly 11.9 million kilograms (26.2 million pounds) of nitrogen and 2.2 million kilograms (4.8 million pounds) of phosphorus per year. How do we get this number? A single dairy cow produces an average of 56.7 kilograms (125 pounds) of manure (both feces and urine) per day, or 20,700 kilograms (45,600 pounds) per year. Thus, 110,000 dairy cows will produce some $110,000 \times 56.7 \times 365 = 2.3$ billion kilograms of manure per year. A single dairy cow excretes 108 kilograms of (238.2 pounds) of nitrogen and 19.7 kilograms (43.5 pounds) of phosphorus (Van Horn *et al.*, 1994, Buckley & Makortoff, 2004; Koelsch 2018;). Again, just using simple mathematics we estimate that these 110,000 cows produce 11.4 million kilograms (26.2 million pounds) of nitrogen and 2.2 million kilograms (4.8 million pounds) of phosphorus per year that must be managed in a sustainable way. In other words, disposal must not significantly disturb the carbon-nitrogen-phosphorus balance of soils in the watershed to avoid offsite loadings into streams and rivers with consequent negative impacts to air and water resources. Is this a lot? To answer this question, we pose the "so what" test: how does this amount of nitrogen and phosphorus translate into an implied application rate (units per area) on the available cropland? And how does that rate compare to typical nitrogen and phosphorus application rates?

To calculate the implied application rate, we need to know the number of hectares (acres) of cropland in Erath County. We do not need a precise number for the purpose of this exercise, a rough estimate will suffice. If we go to the United States Census of Agriculture (National Agricultural Statistics Service, 2019a, b), we find out that the number of farm hectares (acres) in Erath County is about 243,000 hectares (600,000 acres) and one-fifth or 48,500 hectares (120,000 acres) are cropland, although this can vary slightly from year to year. In 2002, 34,000 hectares (84,000 acres) were in field crops suitable for spreading manure (74,000 acres of hay or forage crops, 4,000 acres of corn and sorghum, and 6,000 acres in wheat) (National Agricultural Statistics Service, 2020). We divide the quantity of nitrogen and phosphorus in manure by cropland area to estimate the implied application rate. If the implied application rate far exceeds the typical amount applied for crops, we can reasonably assume that we have a problem with excess nutrients. To simplify the analysis and to give a conservative estimate, we assume all 48,500 hectares of cropland were planted in hay and forage crops.

We now divide 11.9 million kg of nitrogen and 2.2 million kg of phosphorus by 48,500 hectares of cropland. Rounding down the result gives us 245 kilograms per hectare (218 pounds per acre) of nitrogen and roughly 44 kilograms per hectare (40 pounds per acre) of phosphorus. According to Texas A&M Agricultural Extension, hay can take up to 25 kilograms of nitrogen per metric ton (50 pounds per U.S. ton) from all sources depending on the yield and prices (that includes what it gets from rain, legumes, naturally available from organic matter of soils, and, of course, manure) (Perkins, 2018; Wortmann *et al.*, 2013; National Agricultural Statistics Service, 2020; Stichler & McFarland, 2020). In Erath County hay annually yielded, between the years 2002 and 2017, an approximate average of 5.6 metric tons per hectare (2.5 U.S. tons per acre) (National Agricultural Statistics Service, 2019a. Thus, one hectare of hay in Erath county can utilize 140 kilograms of nitrogen [140 = 5.6 metric tons per year × 25 kilograms] (125 pounds per acre [125 = 2.5 tons per year × 50 pounds per U.S. ton]). Comparing our estimate 244 kg versus 140 kilograms (218 pounds versus 125 pounds) of nitrogen suggests that the answer to our "so what" test is: 'We have a problem with nitrogen.' In other words, our test threshold is exceeded under somewhat unrealistically favorable (for the dairy operations) assumptions that farmers growing crops will be able and willing to accept all of the manure that is made available from the livestock operations.

It is important to bear in mind that these estimates are extremely optimistic. First, we assumed all of the cropland could be used to spread the manure as fertilizer, but only about two thirds of the cropland were suitable for applying manure. And importantly as we discussed earlier, the ratio of nitrogen and phosphorus in manure is rarely what is needed by the crop. Generally, manure when applied at a rate appropriate for nitrogen results in an over application of phosphorus. Also, even if the crop could use all of the nutrients, not all of the manure can be collected and used for fertilizer. Even if all manure were collected, some portion of the nitrogen would escape as ammonia. Moreover, the manure would have to be stored properly to prevent its premature release into the environment. Unfortunately, history suggests that storage and transport of this amount of material leads to many spills and unintended discharges onto streams and rivers, let alone emissions into the air in the form of ammonia, nitrous oxide, and methane. These qualifiers serve to highlight the issue Erath County is addressing.

In 1992, USDA estimated that 309 kilometers (192 miles) of the Bosque River watershed were contaminated with livestock pollution. The pollution was not just from excess nitrogen and phosphorus, but also from undesirable microbial populations. This is because the nitrogen and phosphorus in manure, when mixed in with carbon from the straw used for livestock bedding, creates an ideal environment for the growth of microorganisms that degrade water quality for both human and wildlife use. Air quality can be impaired by emissions of ammonia and dust, which not only represents an insult to the senses but also causes degradation of wildlife habitat downwind (Mukhtar & Auvermann, 2020). The amount of nitrous oxide emitted, though not officially documented, can be relatively easily calculated and is undoubtedly substantial. This makes the livestock operations significant contributors to atmospheric emissions of a major greenhouse gas.

7 2.6.1 Take home message

Doing the mathematics, such as the simple mental walkthrough conducted above, to estimate in advance the livestock population that can be accommodated in the watershed might have protected the natural resources. Using the amount of nitrogen and phosphorus produced by the proposed operations would have provided a clear signal that the cropland within the watershed was not sufficient to handle the amount of manure expected. Nitrogen readily changes into numerous forms that can degrade the environment. Storing and transporting manure that contains significant quantities of reactive nitrogen will inevitably result in portions escaping into the air and water resources. About 5% of total nitrogen applied will transform into ammonia. The exact number depends upon a number of factors including how the manure was applied to the soil, what time of year it occurred, and subsequent management of the soil. However, the figures here and below provide a rough ballpark idea of how much is transforming into which nitrogen forms. Research suggests that some 0.3–1% of the nitrogen in manure from a livestock operation will eventually become nitrous oxide (Thorman *et al.*, 2020). N_2O emissions globally from manure rose 34% from 1990 to 2017. This is largely the influence of large-scale operations, which produce massive amounts of waste that is generally over-applied as fertilizer (Lilliston, 2019). In addition to being a source of environmental loadings of rN, large concentrated livestock operations can be sources of major emissions of methane due in large part to the diet of confined animals. Taking into account the number of animals in a watershed will give environmental managers a reasonable estimate of the potential for excess emissions that could result if precautions are not taken.

Excess nitrogen and phosphorus need to move into the long-term storage pools if environmental degradation is to be avoided. This will not happen if the conditions for conversion to soils do not apply, such as in a climatic zone with low rainfall where there would not be enough carbon from plant production to meet the minimum conditions for soil production. The amounts have to be in the right proportions and be present when soil microorganisms are present to convert the excess nitrogen, phosphorus, and carbon into new soil. Conservation measures, such as slowing the water down with vegetation, cover crops to impede water flow, or wetlands to intercept the water, would be necessary. Otherwise, rainfall can move excess nitrogen and phosphorus into waterways. The water, along with the nitrogen and phosphorus it may contain, will go somewhere. If land is not managed to avoid negative impacts, the quality of natural resources will suffer.

Precautions could and should be adopted before visual and other sensual evidence of an environmental problem becomes pronounced. The best approach to avoid a problem is to anticipate the likely outcomes and respect the balancing of carbon, nitrogen, and phosphorus cycles.

7.3 THE ROD SERLING FACTOR: FOR YOUR CONSIDERATION

Here are three exercises for you to apply what you have learned on real environmental issues. These are issues that have a long legacy, requiring a sustained effort to address. Know at the beginning that they pose multiple challenges to their resolution. These include legal, economic, and social barriers. Each proposed solution will touch on at least one of these barriers. To date no painless solution has been identified. You should not expect to identify the 'answer', but the exercises should stimulate thought and promote discussion. Each issue provides you with insight into the interconnectedness that characterizes our environment.

We encourage you to review the background material in Section 7.2 and to revisit Chapters 5 and 6 as needed while doing these exercises.

7.3.1 Greenhouse gas and climate change: what will happen

Some six million years ago, conditions on earth were such that humans emerged onto the evolutionary tree (Harari, 2015). CO_2 levels in the atmosphere were about around 400 ppm (parts per million) (Concio 2019). Over the succeeding millions of years during which human species evolved and thrived, these levels declined to 260–270 ppm (Wigley, 1983.) where they stayed for a million years or more (National Space and Atmospheric Administration, 2020). Earth's atmospheric CO_2 level now stands again at 415 ppm. In two hundred years, the concentration has increased by some 150 ppm or around 60%.

How did this happen? In 2019, 136,762 TWh (Tera Watt hours equivalent) of fossil fuels were burned. This is an increase from 40,553 TWh in 1965 (British Petroleum, 2020). Coal, petroleum, and gas account for the fossil fuels (i.e., old carbon) burned – 14 billion tons of coal, 12 billion tons of oil, and 8 billion tons of gas (Ritchie, 2017). About half of the carbon dioxide produced from the burning of these fossil fuels remained in the atmosphere, the rest was absorbed by the oceans and terrestrial soils and plants. With the increase in CO_2 absorbed by the oceans, ocean acidity has increased by 30% over preindustrial levels (National Oceanic and Atmospheric Administration, 2020).

CO_2 is not the only carbon-containing greenhouse gas that has increased in the atmosphere over the past couple of centuries. Methane concentrations have reached two and a half times preindustrial levels (Jackson *et al.*, 2020). 60% came from agricultural activities before the year 2010 (World Meteorological Organization, 2010) more probably comes from energy production today. And a gaseous form of reactive nitrogen has also contributed to the increase in greenhouse gas levels – nitrous oxide. Nitrous oxide is also increasing at logarithmic rates with levels now 123% higher than in preindustrial times. 40% comes from anthropogenic sources.

Over the past two hundred years, we have shifted massive amounts of carbon and nitrogen from the long-term storage pools of these elements from the earth's crust or the atmosphere (in the case of N) to the relatively short-term pool of the atmosphere where they affect the rates of change in ecosystem processes. This has occurred so quickly that earth's natural processes can only very slowly readjust to new equilibrium levels and at which carbon, nitrogen, and phosphorus strive to come into new balance in soils and the ocean.

Should we be concerned? If we like the conditions that existed millions of years ago and have not existed for nearly the entire existence of humans on earth, then okay. However, the surface warming caused by the greenhouse gases causes more water to evaporate which leads to an increase in atmospheric water vapor (humidity). Remember the mugginess after summer thunderstorms. Since water is also a greenhouse gas, we now have what is called a positive feedback loop with ever more warming.

How will the terrestrial pools of carbon, nitrogen, and phosphorus react? Will there be any effects on sea level rise? Food production? Health? Why?

What would you do to curtail these increases in greenhouse gas levels?

7.3.2 Marginal land and soil erosion

The Homestead and Railroad Acts provided incentives for westward expansion and the plowing of the plains. Much of the land was rich, characterized by deep fertile soils, but a substantial portion of the land was marginal, able to support farmers in times with good weather and prices, but failing to do so when prices dropped or the weather was bad. Farmers trying to survive cropping these lands often found themselves working as hard as they could, but falling further behind. Many of these soils were also fragile, highly susceptible to erosion. In the case of highly erodible soils, the farmers' struggle to stay solvent can become a cost to their neighbors and downstream communities as the soil is blown away as dust or washed off the land and transported into waterways.

Prior to European settlement, much of the grasslands in the North American Great Plains were characterized by deep soils, protected by a diverse mix of grasses and forbs. This prairie vegetation had root systems that extended deep into the soils. Over the millennia these root systems resulted in a high soil organic content. Plowing exposed the organic matter to oxidation, releasing carbon dioxide but also the associated bioavailable nitrogen and phosphorus, benefiting crop productivity. However, by breaking up grassland vegetation and plowing these fragile lands, settlers also exposed the soils to erosion. This erosion over time would reduce productivity, making marginal soils more so.

With the Homestead and Railroad Acts, the first settlers claimed the land available that was closest to the population centers in the east. Later settlers had to travel further and further to the west to locate unclaimed land. In general, annual precipitation decreases as you travel west from the Mississippi River, so the later homesteaders were forced to settle land in areas with less annual precipitation. Often these lands were in the short-grass prairie where the grass was only a few inches high, but the root systems could be 6 feet deep (Milchunas *et al.*, 2011). These deep rooted systems enabled the native vegetation to access water during the periodic multi-year droughts that characterize the western Great Plains. Disturbing these short-grass prairie systems and replacing them with more shallow rooted crops not only exposed the soils to erosion during the periods between tillage operations and crop crown closure but also during droughts when crops could not be established.

The American Dust Bowl is a prime example of how the conversion of native vegetation to crops can go awry. Pioneers settling in western Kansas, Oklahoma, and Texas, and eastern Colorado and New Mexico reached this region during a period of prolonged above average precipitation. Additionally, high wheat prices after World War I made even marginal cropland profitable, attracting more farmers to the region. Conditions changed dramatically in the early 1930s as the precipitation stopped and prices plummeted. As the rain stopped and the crops failed, the constant prairie winds started to blow the unprotected soil (Figure 6.7). The slow-moving disaster of farmers going bankrupt, communities picking up and leaving, and huge regionwide dust storms lead to a federally directed effort to stabilize the soil. It also led to the creation of the Soil Conservation Service (SCS), the predecessor to Natural Resources Conservation Service), as the lead conservation agency in the United States.

SCS developed strategies to protect and stabilize the soil. Many of these practices were discussed earlier in Chapter 6. The primary strategy was to establish vegetation to protect the soil. The lack of precipitation made establishing these covers very difficult. An important outcome from the Dust Bowl and the efforts to stabilize the soils was the recognition that in those areas where the native vegetation remained, the soil did not blow.

Today numerous conservation innovations have helped reduce erosion on these marginal highly erodible lands. The primary tool is to minimize exposing the soil to wind and water. This is accomplished by maintaining vegetation or crop residue on the soil as much as possible. Much cropland has been restored to grass covers, although not with the diversity of the original prairie. Some of these restored grasslands are enrolled into the Conservation Reserve Program and others are now used for grazing livestock. If the land remains in crops, adopting planting methods that reduce tillage becomes important for reducing soil erosion and the offsite impacts.

7.3.2.1 Summary

The settlement of the land west of the Mississippi River resulted in the conversion of hundreds of millions of acres from native vegetation to croplands. Tens of millions of these acres were both marginal and susceptible to erosion. The elevated levels of erosion that occurred from cropping these acres impaired water and air quality for people living downstream and downwind. Concerted conservation efforts have made much progress in reducing erosion, but there are still many acres of cropland eroding at a rate that cannot be sustained to maintain the productivity of the soil. Additionally, cropping these lands has had a significant detrimental on wildlife populations. As noted above, restoring native grasses is an effective tool in reducing erosion and, if sufficient area is restored, it has the benefit of restoring wildlife habitat for species adversely affected from the loss of native vegetation.

7.3.2.2 The challenge

Identify a strategy to restore the ecosystem services provided by the prairie grasslands. This plan should acknowledge the economic activity generated by crop production, including supporting activities such as machinery, seed, fertilizer, and chemical suppliers, grain elevator operators, grain traders, and local businesses. At a minimum it should also consider the effects on air and water quality, wildlife habitat, land values, soil health, and carbon sequestration. And finally, the analysis should consider the countervailing costs and benefits and their unequal distribution, and what mechanisms could be used to avoid having one group or another bear a disproportionate share of the costs or receive an uneven share of the benefits.

7.3.3 Draining wetlands: everything goes somewhere redux

Between the 1780s and 1980s the wetland area in contiguous 48 states of the United States decreased 53% from 89 million hectares to 42 million hectares (221 million acres to 104 million acres) (Dahl, 1990). Of particular interest for this discussion are the wetland losses for the Corn Belt (Illinois (85%), Indiana (87%), Iowa (89%), Ohio (90%), Missouri (87%)), and adjacent states of Minnesota (42%), North Dakota (49%), South Dakota (35%), and Nebraska (35%). Collectively these major agricultural states provide most of the water from the Upper Mississippi, Ohio, and Missouri Rivers and much of the excess nitrogen and phosphorus flowing into these waterways and Gulf of Mexico. Between 2001 and 2005, the Ohio-Tennessee, Upper Mississippi, and Missouri Rivers accounted for 70% of the flow, 94% of the nitrate, 93% of total nitrogen, and 84% of the total phosphorus the Mississippi River delivered into the Gulf of Mexico (Aulenbach *et al.*, 2007). The draining of wetlands and conversion to agricultural use contributed significantly to impaired water quality within the MRB and Gulf of Mexico and, as we noted in Case 1, contributed to higher flood levels in 1993. The draining of an extensive area of wetlands, a system that processed and stored carbon, nitrogen, and phosphorus, led to installing a system that leaks nutrients into the surrounding environment. And as we have repeatedly stressed – everything goes somewhere, gravity is always at work, and natural systems work to balance one another.

We have already examined the role of changes in the landscape on flooding along the Mississippi River and its tributaries in our first case study. By draining wetlands farmers reduced the capacity of the land to store water, which still had to go somewhere. This accelerated the movement of water over and through the landscape, increasing the peak flow associated with each storm and increasing the damages when the river floods. This is straightforward and easy to see. It is also easy to see why and how the draining of wetlands occurred. Each 4 or 40 hectares (10 or 100 acres) had a small effect that was not observed until the cumulative acreage drained was substantial enough for a large storm to cause flood damage. Even then it took a while to recognize that the higher flooding was caused by draining wetlands. After all, there were other modifications being made: levees, dredging, and new construction to name a few.

The full relationship between draining wetlands and diminished water quality is not intuitive. Yes, people can recognize that ponds around cropland are often green with algae and realize that perhaps some fertilizer was getting into waterways. But that is only part of the story. Wetlands can also denitrify nitrate and convert it to N_2 gas, removing rN from the terrestrial pool and returning it to the atmospheric pool. With respect to phosphorus retention, wetlands are less effective. A possible exception is the phosphorus associated with suspended solids (Reddy *et al.*, 1999). What difference could millions of hectares (acres) of wetlands make for reducing rN in the Mississippi River and the Gulf of Mexico? The answer to this question is everything goes somewhere. We need to calculate how the losses of rN from cropland and the reduced denitrification of rN contribute to the problem of hypoxia in the Gulf of Mexico. We also need to account for how the carbon, nitrogen, and phosphorus ratios change under increasing or decreasing land in wetlands.

Mitsch *et al.* (1999) estimated that restoring and constructing 2–5.3 million hectares (5–13 million acres) of wetlands within the MRB would reduce nitrogen loading to the Gulf of Mexico by 300,000–800,000 metric tons per year. They similarly estimated that restoring 7.7–19.4 million hectares (19–48 million acres) of riparian bottomland hardwood forests would have a similar impact. Subsequent analysis suggests that wetland creation and restoration targeted toward areas receiving water with high nitrate concentrations, particularly from drainage systems, would increase the nitrogen removal rate (Crumpton *et al.*, 2006). This reduces the area of wetlands that would be needed to achieve the same level of nutrient removal (U.S. Environmental Protection Agency Science Advisory Board, 2007).

In the time since these analyses were conducted, no significant progress has been made in restoring freshwater wetlands. The most recent assessments of the status and trends in the United States (Dahl & Allord, 1997; Dahl, 2011) found a net loss of 8,800 hectares (22,000 acres) of freshwater wetlands. This loss was not significantly different from zero, given the statistical error of the survey. The assessment found offsetting losses from forested wetlands (256,000 hectares or 633,000 acres), restoration of freshwater emergent vegetation (116,000 hectares or 287,000 acres), shrub wetlands (73,000 hectares or 180,000 acres), and agricultural shallow ponds (46,5000 hectares or 115,000 acres).

Restoring and constructing wetlands to take advantage of their nutrient cycling and storing functions are not likely to be the salvation for excess nutrients in the Mississippi River system because of the likely high economic cost. Any redress will likely involve wetlands to much smaller extent than was envisioned by Mitsch *et al.* (1999). Mitsch *et al.* (1999) identified several other practices that would reduce nitrogen loading in the MRB. These included changes in farm practices including nitrogen management (900,000–1,400,000 metric tons/yr), substituting perennial grasses for 10% of corn–soybean rotations (500,000 metric tons/yr), and improved animal manure management (500,000 metric tons/yr.). Evidence on the adoption of these practices since 2007 is mixed and at times anecdotal. On the discouraging side is the 5.3 million hectares (13 million acres) increase in average corn and soybean planted acres since 2007 (National Agricultural Statistics Service, 2020), and there has been a substantial increase in tile drainage installation and renovation in the last decade (Agricultural Drainage Management Coalition, 2020). Both the increases in corn–soybean planted acres and tile drainage installation will increase nitrogen loss to waterways. Partially offsetting these increases is evidence of an increase in the adoption of nutrient management and manure management practices (unpublished Natural Resources Conservation Service data, 2020b).

7.3.3.1 Take home message

The loss of wetland nutrient cycling and storage function has led to a diminished capacity to process excess reactive nitrogen at a time when there is an increase in rN within the Mississippi River land–water system. Remember if you add more rN to a system, maintaining balance in the system requires removal (denitrification or long-term fixation such as in deep ocean or the more complex fraction of soils) of a commensurate amount. Proposed methods to replace the loss of this wetland function have either not been adopted or have been insufficient to compensate for this loss. If improving water quality within the MRB and reducing the size of the Gulf of Mexico hypoxic zone remain objectives of national policy, either a renewed effort with more resources will be needed or new strategies need to be adopted. Addressing the excess rN in the MRB will take many years. Environmental managers will be at the forefront of these efforts.

7.3.3.2 The challenge

Identify a strategy to restore the ecosystem services provided by the wetlands in the Corn Belt and adjacent states. This plan should acknowledge the economic activity generated by crop production, including supporting activities such as machinery, seed, fertilizer, and chemical suppliers, grain elevator operators, grain traders, and local businesses. It should also include the costs associated with restoring wetlands. At a minimum it should also consider the effects on air and water quality, wildlife habitat, land values, soil health, and carbon sequestration. Attention should be given to differences in addressing excess rN and bioavailable phosphorus, particularly the issue of legacy deposits of phosphorus in river and stream banks and channels. And finally, the analysis should consider the countervailing costs and benefits and their unequal distribution and what mechanisms could be used to avoid having one group or another bear a disproportionate share of the costs or receive an uneven share of the benefits.

Chapter 8
The answer to what is next, summary, and conclusions

We are in the Anthropocene Epoch. A period of time where humans, as much as nature, influence the fate of our planet.

—National Geographic, 2020

We have given you some insight as to how we humans are changing our world. You should know that our decisions about how we manage our resources affect our ability to feed ourselves, the quality of our air and water, and the biological world that surrounds us. In short, how we manage our resources affects our quality of life. This is the meaning of the designation of this period in earth's history as the Anthropocene Epoch. We humans can determine its fate. The responsibility falls upon us to manage its future.

In the case studies we presented to you in the previous chapter, we showed how if the management of natural resources is not well thought out and the scientific principles are ignored, the result will be undesirable consequences. We also showed you how to use core principles and tools to avoid these consequences. In this next section, we provide a brief introduction to public policy which is employed when individual action does not suffice to correct an environmental or natural resource problem.

8.1 THE ANSWER TO 'WHAT NEXT?': PUBLIC POLICY – WHEN INDIVIDUAL ACTION MAY NOT BE ENOUGH

The environmental concerns we have discussed: water quality, air quality, flooding, and climate change to name a few are all the result of numerous agents making

© IWA Publishing 2021. A Guide to Understanding the Fundamental Principles of Environmental Management. It Ain't Magic: Everything Goes Somewhere
Authors: Andy Manale and Skip Hyberg
doi: 10.2166/9781789060997_0149

independent decisions and actions that, in aggregate, degrade a shared resource. Each are examples of the Tragedy of the Commons. Because the services these resources provide are shared, these resources are considered public resources, and public policy is needed to maintain them.

What is public policy? It is the rules we use to allocate and manage public resources. It is needed because coordinated effort, rather than random acts, is required for effective and efficient management of shared resources. Addressing resource degradation in a piecemeal fashion is unlikely to resolve the diminishment of the resource. Consider air, clearly a public resource. You cannot exclude others from using air, but your actions can diminish the benefit for others. Think of a smokestack, spewing noxious gases, reducing the air quality downwind. Similarly, water is a public resource that may be diverted temporarily for personal use, but eventually it moves along within its cycle: evaporating, transpiring, flowing as a river, stream, or infiltrating underground. We share this water with our fellow humans and with the animal and plant worlds. Actions that increase the amount going into the atmosphere diminish the amount available for drinking water and to aquatic ecosystems. Measures that degrade water quality harm those living downstream. Land and soil also provide services that are public resources. They may be privately owned, but their connection to air, water, and biological communities means how they are used affects others.

Public policy has developed over the years in recognition that when resource management decisions affect many people, the interests of those affected need to be taken into account, and only a concerted effort can protect the resource. Managing shared resources to maintain the flow of public services requires that priorities and objectives be identified, strategies to obtain public objectives formulated, and the mechanisms to motivate participants to protect these resources designed. Public policy is the arena where these actions take place.

Because the problems associated with shared resources are complex and different agents are affected differently, how we make decisions and who makes them are critical components of resource management. Addressing these concerns is not easy. Invariably, there will be costs and these costs, as well as the benefits, will not be evenly distributed. Identifying strategies that resolve these problems requires cooperative action on the part of the public and private entities affected. At its heart, is the acknowledgment of the shared resource and the need for a shared response.

It is not necessary to describe here all the forms public policy can take. That is for another text. What is important is recognizing the need for coordinated action based on a science-based strategy to address environmental management concerns. This strategy should incorporate as much information as is feasible. This includes the basic principles and tools we have discussed throughout the text, highlighting the role of the environmental manager in forming public natural resource policy – providing unbiased science-based information to the decision-making body.

8.1.1 Policy affects behavior

Motivating people to adopt the public policy takes three broad approaches – information, incentives, and regulation. Providing information is the least intrusive because it can help lead to better decisions by identifying the costs and benefits of an action. Individuals and organizations can then choose a course of action that maximizes their welfare. An effective tool for encouraging conservation adoption in the United States is the provision of technical assistance to farmers and ranchers interested in addressing resource concerns. The Natural Resource Conservation Service (NRCS) of the United States Department of Agriculture (USDA) or its public and private partners, provides technical assistance in developing a plan that addresses the landowner's resource concern(s), identifies the practice(s) needed, and assures the practice(s) is designed properly. Technical assistance reduces uncertainty for farmers while providing quality control on the planning and design of conservation practices.

Labeling is another means to provide information. It can be used to provide warnings, guidance for proper use, and details of package contents. Perhaps, the most familiar example is the use of warnings to discourage smoking. Labels, such as the nutrition labels on food packages, are also used to provide information on product content and are a tool we all use to obtain information. Packages, cans, and boxes all have the product weight listed on the label, and if they are food product details on the calorie and nutritional content. The information on many labels is a result of policies requiring that specific criteria for determining and displaying information are met.

Labeling plays a role in the environmental issues we have been discussing. Pesticides are required to have labels providing detailed information on product content, how and when a pesticide can and cannot be used, application rates, and medical warnings and guidance. Chemical fertilizers are required to identify the nutrient content and chemical composition of the nutrients on each package. Similarly, most seed packages list the percent germination of live seed from test plots and the percent inert material and weed seed. Both fertilizer and seed companies are required to meet testing standards for the labeling information.

Providing information can help reach public objectives by identifying the consequences of an action. Strategies that reinforce the information provided by offering a realistic alternative can be more effective than relying solely on a communication strategy. Examples include providing dog waste bags and cans next to 'clean up after your dog' signs, or warning labels and application instructions on pesticide containers.

Incentives encourage the adoption of a public policy by reducing the cost of adopting an action consistent with the policy. In conservation, there are numerous state and federal programs that subsidize the adoption of specified practices. These subsidies reduce or cover the cost of installing practices that further the goals of the policy. Disincentives can be used to discourage actions that run counter to policy. These can include fines and fees.

Regulation can also be used to influence behavior. It can range from complete prohibition, to zoning, permitting, and licensing. Frequently, regulation is combined with incentives and information mechanisms to reduce the burden from the regulation, while assuring the policy has its intended outcome. Regulatory oversight often requires certain information to be provided. An example of a combined regulatory–incentive approach would be to encourage the adoption of a practice, say dams, by providing technical assistance, yet requiring permits to assure that incorrectly designed and installed dams do not cause downstream damages. Though the technical assistance reduces the costs of design, it still requires that the landowner go through the time and trouble to obtain a permit.

An emerging approach is the use of private markets to provide incentives for individuals and organizations to adopt practices that address environmental and natural resource concerns (US EPA, 2020c). Manale, (2010) and (Manale et al, (2011) show how markets can address sticky environmental problems. Governments serve the essential role of establishing the baseline conditions and necessary regulatory environment for the market to thrive. Examples of this are defining the commodity, setting minimum environmental standards, delineation of property rights regarding who owns the environmental commodity, and enforcement of contract provisions that serve not just those directly involved in the trades but also the public interest and the environment. Government can act as a neutral party ensuring compliance with contract terms. Polluters can reduce the cost of meeting environmental standards by paying more efficient organizations to make reductions in adverse emissions and for creating environmental offsets, such as through the establishment or restoration of wetlands.

8.2 ADAPTIVE MANAGEMENT – PLANNING UNDER UNCERTAINTY

One particular obstacle for public environmental policy is the incomplete understanding we have of environmental systems. It is also important to recognize that human activity has created imbalances in global systems that have not been previously observed. The U.S. Geological Survey, which collects data on and researches our physical and biological world, calls this phenomenon "loss of stationarity." In other words, our predictive models based on historical data may no longer be valid. With the incomplete understanding of our environment and the loss of stationarity, there are situations where we can no longer take the past and extrapolate it into the future to forecast how systems will react to a plan (e.g. one might not be able to predict the accurately weather let alone the future climate). So how do we proceed?

A prudent course of action is to use adaptive management when implementing a strategy. Forecasts of the likely outcome from implementing a strategy, when the response of a system may have been altered from historical patterns, suffer from greater uncertainty. Adaptive management means developing a strategy to reach a goal based on the best information and science available, monitoring the results from implementing the plan, and modifying the strategy if the outcome deviates from the predicted result. Adaptive management replaces assumptions of predictability with a recognition of the connectedness and interdependence of processes and with an acknowledgment that the plan is likely to require adjustment. This change in the approach has the planners ask several questions when they develop a strategy. Does an action, or public policy, add or detract from balance? Is the impact of the action confined to a small area or to a large region? Can you mitigate function or mimic nature? Can you walk back action? An answer indicating that a strategy will be difficult to revise suggests that the strategy should be reexamined.

8.3 DEVELOPING PUBLIC POLICY

Developing public policy requires the identification of the resource concern, underlying cause(s), objective(s), parties affected, budget, and data needed. These can be used to develop a plan that effectively and hopefully efficiently reaches the stated goal. Identifying the factors generating the concern, the steps needed to resolve the concern, and the parties affected are each an important component in designing a plan. The principles and tools we have presented in this text can be applied in addressing each of these components. Although these principles alone are not sufficient to develop a plan, using them is necessary. That said, they can be effectively employed to demonstrate that a flawed proposed plan will not succeed.

8.4 LESSON SUMMARY

The lessons in this book are:

- Everything goes somewhere. If you do not see where something is going, look harder.
- Gravity is unrelenting. One ignores its effects on water at one's peril. Related to the downward pull of gravity on water is how modifications to the landscape change the speed at which water moves. This leads us to examine the effect of changes in a landscape on water flow and fate. Related to this is consideration of the flow of water other than on the surface, be it by evapotranspiration, subsurface flows, or infiltration. Plans that overlook any of these aspects of water movement are subject to error and unwelcome surprises.
- One needs to observe, measure, and test the assumptions about what is driving change within a system. Environmental systems are complicated with many moving parts. Observing and measuring how these systems react to change are important tools for understanding the reaction to that change. Testing

assumptions and hypotheses can detect what parts of the system need to be better understood.

- Carbon, nitrogen, and phosphorus are the basic chemical elements that make up the building blocks that constitute living things. These chemical elements may change form or even physical state, but their mass is always conserved.
- Water, carbon, nitrogen, and phosphorus each have cycles.
- The carbon, nitrogen, and phosphorus cycles are interrelated and work through atmospheric, soil, terrestrial, ocean, and geochemical processes to maintain a long-term balance within the different pools for each element. These cycles operate at different temporal scales, meaning that an action that creates an imbalance in one may take decades, centuries, millennia, or longer to restore balance. In changing the relationship between long-term and short-term storage of carbon, nitrogen, and phosphorus, we alter their balance in short- and longer-term pools.
- Disregarding the scientific principle of the conservation of mass and altering the mass in pools and the flow of material from one pool to the next have consequences. Understand how actions may alter these cycles by following their paths. Again, if you do not see where something is going, look harder.
- Soils and their health matter. It is in soils that critical components of the carbon, nitrogen, and phosphorus cycles occur and are regulated. If we reduce the carbon content of soils by mismanagement or excessive tilling, we simultaneously reduce the content in soils of nitrogen and phosphorus. The nitrogen and phosphorus will transform. Actions to protect and improve soils can help restore equilibria.
- The rising atmospheric levels of carbon dioxide, methane, nitrous oxide, and other greenhouse gases are a predictable result of human use of fossil fuels, land, and other resources. These elevated greenhouse gas levels have consequences that include higher temperatures and more intense and variable precipitation. These lead to changes in meteorological patterns associated with climate change, increased ocean acidity, and altered biological production.
- Human innovation has removed constraints on reactive nitrogen and phosphorus, increasing agricultural and industrial production, but at a cost. Algae blooms and marine dead zones result from too much reactive nitrogen and bioavailable phosphorus in water bodies. Managing and conserving soils (agricultural, wetland, and forest) are critical to ameliorating these excesses by providing longer-term storage sinks.

8.5 CONCLUSION

We hope you use this introduction to environmental management as a foundation for your future studies and apply these core principles and tools in your endeavors. We also hope it has helped you identify areas for future study as you construct your curriculum. For those who deal with policy or the management of natural resources, we have given you a framework for evaluating proposed solutions to environmental and natural resource problems.

The persistence of sticky environmental issues should serve as a signal that disturbing (changing) these relationships matter. Altering water pathways without following the path of water through its entire cycle is a recipe for long-term problems. Managing the carbon, nitrogen, and phosphorus pools; the flows between and among these pools; and the balancing of cycles lies at the core of environmental management.

And if you are concerned about a changing climate, ask whether or not we are shifting carbon or nitrogen from longer-term storage pools to short-term pools, making them bioavailable and eventually into greenhouse gases. And as a corollary to this, ask whether or not what we are doing in our economic activity can be mitigated through how we manage our natural resources, especially soils.

Remember that sustainable management means that what we have introduced into shorter-term pools is balanced by what is moved into longer-term storage. The chemical elements of life, whether or not they are carbon, nitrogen, or phosphorus, are never destroyed; they can only be managed. How we manage these substances affects the quality of our natural resources and our lives.

Appendix A
Acidity

Water (Figure A.1), as has been explained, is a molecule consisting of two different atoms: two hydrogen atoms and one oxygen atom. Because of the way that the hydrogen atoms are bonded to the oxygen atom – not in a linear or straight line manner but rather at an angle, there is an overall charge to the molecule. That is, the electrons (the small negatively charged particles associated with atoms) are drawn to the oxygen atom, giving the molecule a negative charge adjacent the oxygen atom, and a corresponding positive charge adjacent to the hydrogen atoms where it is now slightly electron deficient. The result is that water molecules tend to attract each other, with the negative pole of the oxygen atom in contact with the positive pole of the hydrogen atom. This attraction is shown in Figure A.2. The attraction between the hydrogen and oxygen atoms holding the water molecules together is a weak bond (or hydrogen bond) compared to the stronger (covalent) bond linking the atoms in the molecule. Nevertheless, in

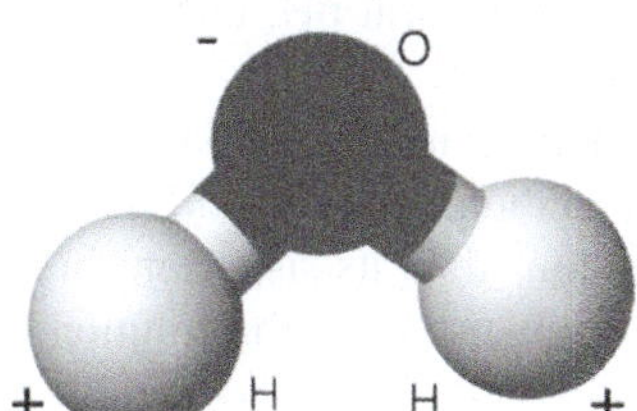

Figure A.1 Water molecule with charges. *Image source*: Wikicommons.

© IWA Publishing 2021. A Guide to Understanding the Fundamental Principles of Environmental Management. It Ain't Magic: Everything Goes Somewhere
Authors: Andy Manale and Skip Hyberg
doi: 10.2166/9781789060997_0157

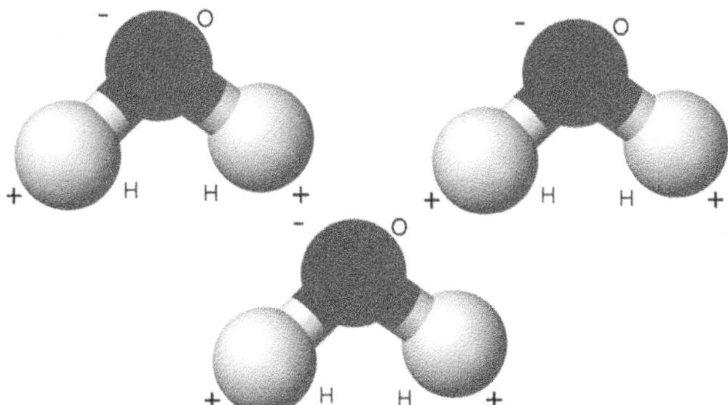

Figure A.2 Water molecules. *Attraction image source*: Wikicommons.

nature, there is always the probability, though a small one, that a hydrogen atom from one water molecule will detach, even for a very short period of time, and bind to the oxygen atom. This hydrogen atom which loses its electron is called a proton. We indicate the deficiency of one electron by attaching the plus sign to the H. In Figure A.3, we show this in a slightly different way using the denotations for oxygen and hydrogen atoms. The dotted line represents the unstable, temporary bond between the hydrogen **ion** (a charged atom or molecule) and the oxygen in the water molecule and the solid lines the stable, covalent bonds between oxygen and hydrogen in water. The resulting **hydronium** ion is positively charged. The now lonely OH ion has a negative charge, indicating that it has an extra electron. The probability of this occurrence in water can be readily calculated. In a glass of water, the concentration will be 1.0×10^{-7} **moles** per liter. This is a pretty small number. Chemists display this number more simply as $pH = 7$ using the inverse of the **logarithm** of the number. In pure water, OH ion will have the same concentration as H^{+}. Thus, the glass of water is said to have a neutral pH, that is the water the concentration of H^{+} equals the concentration of OH^{-}.

In itself, this phenomenon is not particularly interesting. It becomes more interesting when the concentrations of H^{+} and OH^{-} are not equal – when H^{+} is present at greater concentrations than its sister ion. In this case, scientists say that the solution is acidic. In the obverse case, the solution would be basic or alkaline.

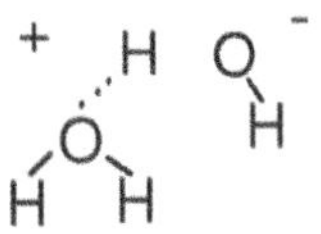

Figure A.3 Water and H ion.

A glass of vinegar has a pH value of roughly 2.4. The hydronium concentration is about 4×10^{-3} moles per liter or 0.004 mol/l. Baking soda is an example of a basic solution with a pH value of 9.

Acids and bases are important in nature because excess charge, positive or negative, can react with other chemicals, thus breaking apart otherwise stable compounds. Carbon dioxide, for example, has a pH value of 5.6 and thus is slightly acidic, that is, a weak acid. High concentrations of carbon dioxide in the oceans lead to a lowering of its average pH value which otherwise is around 8.1. As more CO_2 is absorbed, the ocean pH declines, creating difficult conditions for organisms like clams and mussels and certain types of corals. The lower pH breaks apart calcium carbonate, making it unavailable for shell making.

Appendix B
Chemical elements of life

Of possible interest to our readers is the story of why we focus on carbon, nitrogen, phosphorus, and water to tell the inside story of natural resource management. Is not life made up of more chemical elements than these? And more basically, why is life composed of these chemical elements and not others?

In this discourse, we can only provide a glimpse into an emerging and blossoming area of scientific research that is scarcely more than a hundred years old and has risen in academic and scientific interest since the 1960s when we began to explore space. Scientists, in recent years, can do more than just ponder the question of whether or not life exists elsewhere in the universe. Are the circumstances that led to life on earth, unique? A corollary to this question is what were the circumstances that led not just to life on earth, but conditions that enabled the rise of *Homo sapiens* – us, which should be of relevance to all of us? How repeatable are they? Are our natural resource systems resilient?

These questions have particular relevance to the discussion of natural resource management because, by exploring these questions, we can shed light on what the tipping points are in our systems and what are the elements of most relevance to the survival of our ecosystems. If we cause significant harm to our environment through pollution or modification of the landscape, draining wetlands or straightening rivers, for example, will the system recover? What must humans do to facilitate this recovery?

If we both physically and chemically modify our air, our water, and our soils, will our environment revert to its initial condition if we stop the insult? Can we even

© IWA Publishing 2021. A Guide to Understanding the Fundamental Principles of Environmental Management. It Ain't Magic: Everything Goes Somewhere
Authors: Andy Manale and Skip Hyberg
doi: 10.2166/9781789060997_0161

hope to restore ecosystems that have been degraded? These questions are ongoing in the scientific community and are at the core of environmental and natural resource management. This book helps you begin to ask the right questions and to point you in the direction of where you can seek more information and understanding.

The science of ecosystems has been informed in recent years by the new science of astrobiology and in connection to this, the new science of **prebiotic** chemistry, a field of inquiry that scarcely existed when the authors were in school. We provide you here with the barest of introductions.

Early in the twentieth century, scholars identified carbon, hydrogen, nitrogen, oxygen, phosphorus, and sulfur as the key chemical (biogenic) elements of life (Schwartz, 2006). Over time, scholars began to ask why these chemical elements and not others. This is a particularly important question for astrobiologists who seek to determine not just whether or not life exists outside of earth, but also in what form. Hydrogen and oxygen, when combined, constitute water. Water has the important quality of expanding when frozen, thus becoming lighter by unit mass and preventing oceans from freezing from the bottom. The constant melting and freezing of ice splits even the toughest of rocks, leading to weathering and exposing small chemical units within large rock masses. This frees them up for interaction with other essential elements. For a discussion of the other attributes of water that makes it perfect for life, see Henderson (1913). The elements such as hydrogen, oxygen, carbon, and nitrogen make up 99% of living tissue. They are also the smallest and abundant elements in the universe with stable energy states (electron configurations). If you want to build a stable structure, you want building blocks that can have more than just one point of attachment. Other than hydrogen, these elements do have more one than potential bond. Later researchers included phosphorus and sulfur in this list because they met the above criteria and also can form multiple bonds. Additionally, their bonds tend to be weaker, allowing for easier energy transfer – an attribute necessary to provide and temporarily to store energy for biochemical processes in living organisms.

To start a fire you need kindling, which is carbon. When you start the fire, you blow on the flame. You may not have thought about this, but what you are doing is adding extra oxygen, which while forming new bonds with the carbon spontaneously releases energy. And voila, the kindling burns and heat is released. The oxygen found in the air we breathe is in its gaseous form exists as a molecule consisting of two atoms, with a tendency to combine with other elements. In other words, it is reactive.

For organisms, oxygen bonding to carbon releases energy which is taken up and stored by molecules in the cell. Under certain circumstances, sulfur can serve this role and appears to do so in deep oceanic vents where oxygen is scarce. Much earlier in earth's history the atmosphere was very different, the oxygen molecule O_2, the 21% of the atmosphere that we breathe, was unavailable. Because of its reactivity, the oxygen that existed quickly combined with other elements, often in mineral complexes with iron. [Observe how a piece of iron, left uncovered, will over time turn red with rust, especially if made wet. Water is a molecule of oxygen and two hydrogens. What you are observing is the 'oxidation', that is the combining of oxygen with iron.]

So how did things change? How was the oxygen released from its entrapment in mineral complexes? Most theories point to living organisms, but not life as we know it. Life, beginning in the oceans, consumed the oxygen associated with minerals (such as iron minerals) and possibly the sulfur in rocks to generate energy and make the carbon to carbon bonds needed to construct the building blocks of life. Molecular oxygen, the gaseous version that we breathe, was the waste product. The accumulation of this byproduct created, over eons of time, the atmosphere that we know.

Clearly, oxygen is a key natural resource material without which life, as we know it, would abruptly stop. Why then are we giving it short shrift in this book? Because its absence in our atmosphere or oceans would mean our very rapid demise. If we ever get so close to mismanaging our environment that oxygen levels are diminished, then our subject matter will be beyond even academic interest. We focus on what we believe are important matters that, with prudent measures, we can manage and affect; we omit what is too dire to contemplate. Furthermore, if we manage nitrogen, carbon, phosphorus, and water, we also manage oxygen.

What about sulfur, which is also a key component of cells? Like oxygen, sulfur is very abundant in the universe. On earth, it is almost totally locked up in dense mineral complexes. It is particularly reactive, which means, like oxygen, when it is released into our biosphere, it quickly becomes bound into chemical complexes, thereby relatively unavailable for bioreactions. Hence, for the purposes of natural resource management, sulfur is a minor actor. An exception is when it is mined or made available through combustion processes. In these cases, it is available as a local pollutant. [Remember the smell of rotten eggs? What you are smelling, at low concentrations, is hydrogen sulfide.] The sulfur reacts quickly with elements in the surrounding environment. In doing so, it can create a very strong acid, sulfuric acid, that poses a major negative impact on the environment.

We have been introduced to hydrogen earlier, but in context with its familiar partnering with oxygen – water. Although hydrogen plays an important and significant role in living organisms, it is not a major player in resource management when alone, except as affecting acidity. Though it is found throughout the biological world, it only has one binding site. It occurs as an appendage, connected to the other key chemical elements that have more than one site to which other atoms can be bound and affects the folding of molecular chains and thus the 3D structure of molecules. In the biosphere, its scarcity or abundance relates to its association with oxygen in the chemical compound water (H_2O). As we state over and over, balance is everything in nature. Even here, nature provides that excess hydrogen is balanced or buffered so as not to disturb the harmony that enables life for higher, more complex, organisms.

To create the necessary conditions for even the simplest life, early earth had first to lock up the excess H in mineral storage, another process that took billions of years (Krissansen-Totten, 2018). Over time, on land, soils provided the sealant, keeping the biosphere at a balanced or roughly neutral acidity.

Our discussion of the key chemical elements of life should not mislead the reader with the notion that life required only the presence of these elements for self-assembly into the molecular building blocks of life. This is where prebiotic chemistry comes in. Moreover, it is here where the question of **repeatability** and **restorability** comes in. Did the initial assembly depend upon prebiotic conditions? If so, then subsequent assembling becomes nearly impossible because these prebiotic conditions currently exist in few places on earth. Recent research suggests that much assembly through complex and as yet poorly understood abiotic processes had to occur before the building blocks could be constructed. These are processes that scientists estimate took hundreds of millions if not billions of years to occur.

Just one example should serve to illustrate this point – how phosphorus, which most likely began its existence on earth in mineral form combined with oxygen, which as we have indicated above, was in its nonmolecular form scarce, to form phosphate (Schwartz, 2006). For phosphorus to function as a key component in life's building blocks, it had to be freed from its mineral complex in rock, generally as **apatite**, commonly seen as a rock outcropping, and combined with oxygen (Schwartz, 2006). Most likely, weathering, volcanic activity, and the complex brewing of chemical mixtures in deep ocean vents freed these elements from their terrestrial graves, and water served to bring these substances together, facilitating the formation of phosphates (Abe, 2001; Schwartz, 2006).

Remember that to be a key ingredient for building blocks of life, the chemical element or molecule (such as NH_4^+) must be present in very large quantities. Thus, an event occurring in a volcanic vent, a lightening-strike or similar event, had to be repeated countless times and result in more of the molecule in stable and in a potentially **bioreactive** state than its rate of decomposition. Although the event (the creation of the new molecule) is rare, over hundreds of millions of years, a rare event happens many times and the new molecule accumulates. Note the importance of the role of time in early earth, slowly and gradually changing the initial chemical state of the elements of life that were present in early earth into molecules that could be reassembled into life's building blocks. The rocks and minerals we see and experience in our hikes and visits to wild lands are product of millions of years of transformation. To explore these prebiotic conditions that no longer exist on earth and to understand the processes that led to the rocks and minerals that may have occurred in early earth, scientists probe asteroids and comets.

Appendix C
Building blocks of life

The building blocks of life, the basic molecules that life uses to build the components of cells, are amino acids, sugars, lipids, and nucleotides.

Amino acids: There are 20 amino acids that comprise the building blocks of proteins and coded for in the genetic code. Each amino acid shares what are called amino and carboxyl groups. They differ by the moiety that is attached to each. Depending upon their sequence, they make different types of proteins for disparate functions. Proteins are used as structural components of cells, connectors of cells, a means for movement, a weapon in the immunological war on microbial invaders, and as a conveyer of information or signaler.

Sugars: Sugars are the simplest group of carbohydrates, best known for the role in storing energy. They can comprise single molecules, such as glucose, or double molecules, such as sucrose, which contains a molecule of glucose and one of fructose. Sugars serve as the fuel for cellular biochemical reactions, as energy storage, as a component of DNA and RNA molecules, and even as signalers in intercellular communication. Different sugars are linked together to form energy storage units, such as starch, or larger structures on the surface of cells for communication and attachment.

Lipids: Lipids are molecules composed of a charged group that can attract water and long carbon-containing chains that repel water. Cell membranes are made up of lipids that, because they repel water, cluster, creating a barrier between the outside water environment and the internal world of the cell. Lipids are the fat in fat cells, serving as energy-storage molecules. They are also the

© IWA Publishing 2021. A Guide to Understanding the Fundamental Principles of Environmental Management. It Ain't Magic: Everything Goes Somewhere
Authors: Andy Manale and Skip Hyberg
doi: 10.2166/9781789060997_00165

basic building blocks for steroids that serve in communication both within the organism and without.

Nucleotides: The nucleotides (five major ones) are the basic structural units of the genetic material of cells. They form long chains, DNA and RNA, the sequence of which stores information, the genetic code, that directs and regulates cells.

Appendix D
Ecosystem services

Humans have known for thousands of years that natural systems – such as soils, forests, wetlands, and grasslands – provide us with food, fiber, and fuel, along with recreation and spiritual soothing. As civilizations have developed and populations have increased, we have converted these ecosystems in ways that benefit humans solely for the short run, such as for crops and food today versus water storage and flood mitigation tomorrow. In recent years, society has become aware that natural disasters are often the product of human intervention in the natural world; the loss of these lands begins to impair or detract from the human built environment.

With expanding research and knowledge into how our natural systems work, scientists have begun to identify and quantify these benefits that they provide. Modern technology, especially computers, Geographic Information Systems, and new sensing technology have allowed us to quantify these benefits, which we call **ecosystem** services. With measurement and quantification, we have an analytical framework for comparing what the land, not just pristine landscapes but also lands altered for human benefit such as agricultural land, provides towards supporting human health and welfare.

The World Health Organization (WHO) in a project in what is referred to as the Millennium Assessment created a roadmap for identifying, characterizing, and quantifying these services. It defines ecosystem services as the 'combined physical and biological components of the environment. These organisms form complex sets of relationships and function as a unit as they interact with their

© IWA Publishing 2021. A Guide to Understanding the Fundamental Principles of Environmental Management. It Ain't Magic: Everything Goes Somewhere
Authors: Andy Manale and Skip Hyberg
doi: 10.2166/9781789060997_0167

physical environment.' It provides a useful taxonomy that combines physical and biological attributes of the landscape to distinguish supporting services that enable nutrient recycling, soil formation, and primary function. These supporting services, in turn, provide for provisioning, regulating, and cultural services. Examples of provisioning services are food, fresh water, and wood for fuel. Those under regulating services are climate, flood, and disease regulation, among others. And for cultural, examples of the services are aesthetic, spiritual, and recreational.

Recommended Resources

BOOKS

Barry J. M. (1997). Rising Tide: The Great Mississippi Flood of 1927 That Changed America. Simon and Shuster, New York, USA.

Leopold A. (1949). The Sand Country Almanac. Penguin, New York, USA.

McPhee J. (1989). The Control of Nature. Noonday Press, New York, USA.

Egan T. (2006). The Worst Hard Time. Houghton Mifflin, Boston, MA, USA.

Wilson E. O. (1992). The Diversity of Life. Norton and Company, New York, USA.

World Commission on Environment and Development. (1987). Our Common Future, also known as The Brundtland Report Oxford University Press. ISBN 019282080X/.

Carson R. (1962). Silent Spring. Houghton Mifflin. New York, USA.

duBuys W. (1999). Salt Dreams: Land & Water in Low-Down California. University of New Mexico Press. Albuquerque, NM, USA.

Reisman M. (1986). Cadillac Desert. Penguin Books. London, UK.

WEBSITES

Soil Science Society of America. www.soils.org

Soil and Water Conservation Society. www.swcs.org

Consortium for Greenhouse Gas Mitigation Strategy, at USDA REEIS. https://portal.nifa.usda.gov/web/crisprojectpages/0189033-consortium-for-agricultural-soils-mitigation-of-greenhouse-gases-casmgs.html

Intergovernmental Panel on Climate Change, IPCC Reports. https://www.ipcc.ch/reports.

MOVIES

Dirt! The Movie (2009). Directed by Bill Benenson, Gene Rosow, Eleonore Dailly. With Jamie Lee Curtis, Bill Logan, Vandana Shiva, Fritjof Capra – IMDb. https://www.imdb.com/title/tt1243971 (20 November 2011).

© IWA Publishing 2021. A Guide to Understanding the Fundamental Principles of Environmental Management. It Ain't Magic: Everything Goes Somewhere
Authors: Andy Manale and Skip Hyberg
doi: 10.2166/9781789060997_0169

Glossary

Acid – Any of various typically water-soluble and sour compounds that in solution are capable of reacting with a base to form a salt, redden litmus, and have a pH less than 7, that are hydrogen-containing molecules or ions able to give up a proton to a base, or that are substances able to accept an unshared pair of electrons from a base.

Acre – 43,560 square feet (4,047 square meters) or 0.4 hectare.

Aerobic – Occurring only in the presence of oxygen.

Aesthetic – Pleasurable to the senses.

Aggregate – Composed of mineral crystals of one or more kinds or of mineral rock fragments.

Agriculture – The science, art, or practice of cultivating the soil, producing crops, and raising livestock and the preparation and marketing of the resulting products.

Agro-environmental – Relating to the impact of agricultural practices on the environment.

Air emissions – The release of various gasses and particles into the atmosphere.

Amino acids – A carbon-based structure that functions as the building blocks for proteins.

Ammonia – (NH_3) A pungent colorless gaseous alkaline compound of nitrogen and hydrogen that is very soluble in water and can easily be condensed to a liquid by cold and pressure.

Anaerobic – Occurring or existing in the absence of oxygen.

Anthropocene Epoch – The period of time during which human activities have had a demonstrable environmental impact on the Earth that is regarded by some as constituting a distinct geological age.

Anthropomorphic – Having human attributes.

Apatite – Any of a group of calcium phosphate minerals that are the chief constituent of phosphate rock.

Aquifer – A water-bearing stratum of permeable rock, sand, or gravel.

Atmospheric pool – Reservoirs of various substances (generally nitrogen, oxygen, argon, and carbon) in the atmosphere which can take or release these substances.

Atom – The smallest particle of an element that can exist either alone or in combination with other atoms.

ATP – ($C_{10}H_{16}N_5O_{13}P_3$) Adenosine triphosphate is a phosphorylated nucleotide composed of adenosine and three phosphate groups that supplies energy for many biochemical cellular processes.

Azospirillum bacteria – A species of bacteria that is able to fix nitrogen in environments lacking oxygen.

Base – Compounds that in solution that have a pH greater than 7, the opposite of acid.

Base flow – (1) The lateral movement of the water table. (2) The portion of the streamflow that is sustained between precipitation events.

Bedrock – The solid rock underlying unconsolidated surface materials (such as soil).

Bioavailable – The amount of an element or compound that is accessible to an organism for uptake or adsorption across its cellular membrane. Almost exclusively, plant roots and soil organisms uptake contaminants that are dissolved in water.

Biogeochemical pools – Reservoirs of substances in the Earth's crust, and oceans, soils, and terrestrial matter, and the atmosphere that have the capacity to both take in and release substances.

Biology – A branch of knowledge that deals with living organisms and vital processes.

Biomass – The amount of living matter (as in a unit area or volume of habitat).

Bioreactive – Biologically or biochemically reactive.

Bioreactors – A vessel that carries out a biological reaction and is used to culture anaerobic cells for conducting cellular or enzymatic immobilization. In conservation bioreactors a designed to promote denitrification of nitrate-rich water.

Biosphere – The part of the world in which life can exist.

Biotic – Related to living things, the opposite of abiotic (nonliving).

Canopy – The uppermost spreading leafy or branchy layer of vegetation.

Canopy interception – The catching or stopping of precipitation by the uppermost spreading leafy or branchy layer of vegetation.

Capillary action – The action by which the surface of a liquid where it is in contact with a solid (as in a capillary tube) is elevated or depressed depending on the relative attraction of the molecules of the liquid for each other and for those of the solid.

Carbon – The nonmetallic chemical element with atomic number 6 that readily forms compounds with many other elements and is a constituent of organic compounds in all known living tissues.

Carbon cycle – The cycle of carbon in the Earth's ecosystems in which carbon dioxide is fixed by photosynthetic organisms to form organic matter, travels through biogeochemical pools, and is ultimately returned to its gaseous state (by respiration, protoplasmic decay, geologic processes, or combustion).

Carbon pools – Reservoirs of carbon that have the ability to release or take in carbon for various and varying periods of time.

Carbonaceous – Relating to, containing, or composed of carbon.

Catchments – An area in the landscape, most often within a watershed, that catches water from precipitation.

Chemical – A substance with a distinct molecular composition that is produced by or used in a chemical process.

Chemical bond – Any of several forces, by which atoms or ions are bound in a molecule.

Chemistry – The science that deals with the composition, structure, and properties of chemical substances and with the transformations that they undergo.

Climate – Average pattern of weather for a particular region over a substantial number of years.

Climate change – Significant and long-lasting change in the Earth's climate and weather patterns.

Colloid – A substance that consists of particles dispersed throughout another substance which are too small for resolution with an ordinary light microscope but are incapable of passing through a semipermeable membrane.

Colloidal particles – The particles that are suspended within a colloidal substance.

Combustion – An unusually rapid chemical process (such as oxidation) that produces heat and usually light.

Conservation – The careful preservation and protection of something, especially the planned management of a natural resource to prevent exploitation, destruction, degradation, or neglect.

Continental shelf – A shallow submarine plain of varying width forming a border to a continent and typically ending in a comparatively steep slope to the deep ocean floor.

Contour farming – The practice of tilling sloped land along lines of equal elevation to conserve rainwater and to reduce soil losses from surface erosion.

Cover crops – Plants that are grown for the purpose of covering the soil to manage soil erosion, soil fertility, soil quality, water, weeds, pests, diseases, biodiversity, and wildlife, and not with the primary purpose of being harvested.

Crop residue – Materials left behind on the soil after a crop is harvested.

Crop rotations – The system of varying successive crops in a definite order on the same area of land, especially to avoid depleting the soil and to control weeds, erosion, and pests and to reduce fertilizer use.

Crustal pool – Reservoirs of various substances (generally carbon, nitrogen, and phosphorus) in the Earth's crust which can take or release these substances.

Dam – A barrier constructed to hold back water and raise its level, forming a reservoir.

Dead zones – Areas with low levels of oxygen that fail to support marine life.

Deforestation – The removal or destruction of forests and conversion to other uses.

Denitrification – Microbially facilitated process where nitrate (NO_3^-) is reduced and ultimately produces molecular nitrogen (N_2) through a series of intermediate gaseous nitrogen oxide products.

Desiccation – Removal or loss of moisture.

Diazotrophs – Single-celled organisms that can convert nitrogen from the air into ammonia.

Dipolar molecules – A molecule that has two ends with opposing positive and negative charges.

Disapparate – To disappear magically.

DNA – Deoxyribonucleic acid, a long-chained molecule, often in the structure of a double helix, that contains the genetic instructions for single- or multi-cell organisms or even some viruses.

Drought – A prolonged period in which there is a lack of precipitation and water accompanied with the presence of dryness.

Dry matter – The mass of a substance when completely dried.

Ecology – The study of organisms, their environments, and the interrelations between them.

Ecosystem – A biological community of interacting organisms and their physical environment.

Element (or chemical element) – The chemical substances that cannot be broken down using chemical reactions and whose atoms all have the same number of protons, often presented in a tabular array called the Periodic Table of Elements.

Environment – The complex of physical, chemical, and biotic factors (such as climate, soil, and living things) that act upon an organism or an ecological community.

Ephemeral stream – A temporary stream that only appears during and directly following a period of precipitation.

Equilibrium – State in which a process and its reverse occur at equal rates whereby no overall change takes place.

Erosion – The gradual geological process in which materials, particularly soil, are worn away and transported by natural forces such as wind or water.

Estuary – The tidal mouth of a river, where the ocean tide meets the stream, home to unique plant and animal communities that have adapted to brackish water – a mixture of fresh water draining from the land and salty seawater.

Eutrophication – Excessive richness of nutrients in a lake or other body of water, frequently due to runoff from the land, causing a dense growth of plant life and subsequent death animal life from lack of oxygen.

Evaporation – The process of turning a liquid into a gaseous vapor.

Evapotranspiration – The process by which water is transferred from the land to the atmosphere by evaporation from the soil and other surfaces and by transpiration from plants.

Fauna – All of the animal life of a particular region, habitat, or geological period.

Feed grains – Cereal grains that used to feed livestock.

Fertilizer – Any substance used to fertilize the soil, especially chemical nutrients, particularly those containing reactive nitrogen or bioavailable phosphorus.

Floodplains – An area of low-lying ground adjacent to a river, formed mainly of river sediments and subject to flooding.

Flood – An overflow of water that submerges land that is usually dry.

Flora – All of the plant life of a particular region, habitat, or geological period.

Fossil fuel – A natural fuel such as coal or gas, formed in the geological past from the remains of living organisms.

Frankia – A genus of bacteria that fix nitrogen.

Fungi – Any of a group of spore-producing organisms feeding on organic matter, including molds, yeast, mushrooms, and toadstools.

Fungicide – A chemical that kills fungi.

Gas – One of the three states of matter, the others being solid and liquid, that has no definite volume or definite shape.

Genetic diversity – The total genetic characteristics in the genetic makeup of a specific species.

Gigaton (or gigatonne) – One billion metric tons (metric ton = 1000 kilograms).

Grass filter – A strip of dense herbaceous vegetation and grass, that filters runoff and removes contaminants before they reach water bodies or water sources, such as wells. Sometimes called grass buffer.

Gravity – The natural force that causes objects to fall towards the Earth due to the physical attraction of the mass of the Earth for bodies at or near its surface.

Greenhouse gas – A gas that absorbs and emits radiant energy within the thermal infrared range that causes heat to be trapped within the Earth's atmosphere.

Groundwater – Water present underground in soil or in pores and crevices in rock.

Guano – The excrement of seabirds and bats, sometimes used as fertilizer.

Hazardous waste – Any industrial by-product, often from the manufacture of chemicals, that is destructive to the environment or dangerous to the health of people or animals.

Heat – The kinetic energy associated with the random motion of the molecules, atoms, or smaller structural units of which matter (solid, liquid, and gas) is composed.

Hectare – 10,000 square meters or 2.471 acres.

Herbicide – A substance that is toxic to plants used for destroying unwanted vegetation.

Humidity – A quantity representing the amount of water vapor in the air.

Humus – The organic component of soil, formed by the decomposition of leaves and other plant material by soil microorganisms.

Hydrologic cycle – See water cycle.

Hydrologist – A scientist who researches the distribution, circulation, and physical properties of the Earth's underground and surface waters.

Hydronium – H_3O^+, a positively charged ionic compound, the concentration of which determines the pH or acidity of a solution.

Hypoxia – A very low level of oxygen, as in an organic environment.

Hypoxic zones – Areas in which there is not adequate oxygen supply often resulting in the death of marine animals and plants.

Ice age – An extended period of reduction of the Earth's temperature, characterized by extensive glaciation and ice sheets.

Inert – A substance that is not reactive under normal conditions.

Infiltration – Is the process by which water on the ground surface enters the soil.

Inorganic – Not consisting of or deriving from living matter and therefore not containing of carbon.

Insecticide – A chemical or substance used to kill insects.

Interception – Precipitation that does not reach the soil, but is instead intercepted by the leaves, branches of plants, and the forest floor.

Intermittent stream – Is a seasonal stream that ceases to flow during seasons.

Ion – An atom or molecule with a net electric charge due to the loss or gain of one or more electrons.

Irrigation – The process of applying controlled amounts of water to plants at needed intervals.

Karst – Area of land consisting of limestone characterized by underground passages, caves, and sin.kholes.

Labile – Easily broken down.

Lateral movement – Water movement along or below the land surface towards a waterway.

Law of conservation of mass – Chemical law that says that the total mass of a chemical element always remains constant, with matter never created or destroyed.

Legumes – A plant that is a member of the pea family, typically associated with fixing nitrogen from the atmosphere.

Leguminous – Relating to or denoting plants of the pea family.

Levee – An embankment built along a river intended to prevent the overflow of a river onto the floodplain.

Lipids – Any of a class of organic compounds that are fatty acids or their derivatives and are insoluble in water but soluble in organic solvents, including many natural oils, waxes, and steroids.

Liquid – A state of matter that has definite volume but not definite shape.

Logarithm – A quantity representing the power to which a fixed number (the base) must be raised to produce a given number.

Macroorganism – A multi-celled organism, often visible with the naked eye.

Manure – Animal waste, can be used for fertilizing crops.

Mass – A property that represents the quantity of matter, particularly in a sample, or the number of atoms of a chemical element involved in a chemical reaction.

Matrix – A surrounding medium or structure.

Microbial – Relating to or characteristic of a microorganism, especially a bacterium.

Micron – A unit of length equal to one thousandth (10^{-3}) of a millimeter or one millionth (10^{-6}) of a meter.

Microorganism – Any organism that is too small to be seen with the naked eye, generally single-celled.

Mineralize – Convert (organic matter) wholly or partly into a mineral or inorganic material or structure.

Mitigation – The reduction in harmful effects of some thing or action.

Mole – A mole, a term used in chemistry, is defined as exactly $6.02214076 \times 10^{23}$ particles and serves as unit for the amount of substance in the International System of Units (SI).

Molecules – A group of atoms bonded together, representing the smallest fundamental unit of a chemical compound that can take part in a chemical reaction.

Natural resource – Materials or substances such as minerals, forests, water, and fertile land that occur in nature and that can be used for personal, social, or economic gain.

Nitrate – NO_3, a nitrogen compound that is a common component in fertilizer and soluble in water.

Nitrogen – The 7th element on the periodic table, that is a gas at room temperature, making up a large percentage of the Earth's atmosphere.

Nitrogen cycle – The biogeochemical cycle by which nitrogen is converted into multiple chemical forms as it circulates among atmosphere, soil, terrestrial, and oceanic pools.

Nitrogen fixation – The chemical processes by which atmospheric nitrogen is assimilated by certain microorganisms into organic compounds, as part of the nitrogen cycle.

Nitrogen pools – Reservoirs of nitrogen which can receive or release nitrogen.

Nonpoint source pollution – Pollution resulting from diffuse sources, generally not traceable to a discrete source, such as a smokestack or drainpipe.

Nucleotides – A compound consisting of a nucleoside linked to a phosphate group, forming the basic structural unit of nucleic acids such as DNA.

Nutrient – A substance that provides nourishment essential for growth and the maintenance of life.

Nutrient cycling – The cyclic movement and exchange of nutrients from organic and inorganic matter into the production of organic matter and back again.

Organic – Relating to or derived from living matter and/or carbon.

Organism – An individual animal, plant, or single-celled life form.

Overland flow – The flow of water occurring along the ground surface when excess rainwater, stormwater, meltwater, or other sources cannot infiltrate into the soil.

Oxidation – The loss of electrons by an atom or a molecule in a chemical reaction often to the chemical element oxygen.

Peak flow – The maximum rate of stream flow during the period of runoff caused by a storm. The highest water level after a storm.

Perennial stream – A stream, river, or channel which flows throughout the year.

Permafrost – A thick subsurface layer of soil that remains frozen throughout the year, occurring chiefly in polar regions of the planet.

Permeability – The state or quality of a material or membrane that allows liquids or gases to pass through it.

pH – A figure expressing the acidity or alkalinity of a substance on a logarithmic scale on which 7 is neutral, with lower values being more acid and higher values more alkaline.

Phosphate – PO_4^{3-} is a chemical moiety (part of another molecule) that contains the chemical element phosphorus, and that is crucial to energy processes in living organisms.

Phosphorus – The chemical element of atomic number 15 that rarely occurs alone but generally in association with oxygen atoms.

Phosphorus cycle – The biogeochemical cycle by which phosphorus is converted into multiple chemical forms as it moves among soil, terrestrial, and oceanic pools.

Phosphorus pools – Reservoirs of phosphorus which can take or release phosphorus.

Photosynthesis – The process by which green plants and some other organisms that contain the pigment chlorophyll capture solar energy to synthesize sugars from carbon dioxide and water.

Physical state – Arrangement of atoms or molecules in matter: solid, liquid, or gas.

Policy – A course or principle of action adopted or proposed by a government, party, business, or individual.

Pollution – The presence in or introduction into the environment of a substance or thing that has harmful or poisonous effects.

Porous – The quality of having minute spaces or holes through which liquid or air may pass.

Prairie – Temperate grassland and savanna ecosystems with productive soils well-suited for crop and livestock production.

Prebiotic – Existing or occurring before the emergence of life.

Protein – Large molecules that are made up of one or more chains of amino acids.

Reactive – The tendency of a chemical to interact with itself or another chemical with an overall release of energy.

Reforestation – The process of replanting an area with trees.

Repeatability – The ability to replicate the results of an experiment.

Reservoir – A large natural or artificial lake used to store and supply water or a pool of a substance or material.

Resource management – Manner in which societies manage the supply of or access to the resources upon which they rely for their survival and development, generally used in reference to, natural resources.

Restorability – Quality of being able to be restored or reclaimed.

Retention – The continued possession, use, or control of something.

Rhizobia – Bacteria that fix nitrogen inside root nodules of legumes.

Riparian forest buffer – A forested area adjacent to a stream, which intercepts and filters runoff from upland lands and helps shade the stream.

rN – reactive nitrogen.

Rod Serling – An American writer and television producer known for the television series, The Twilight Zone. Said, "Offered for you consideration."

Root nodules – A part of the roots of a plant, often legumes, that form symbiosis with nitrogen-fixing microorganisms.

Runoff – The flow of water occurring on the ground surface when excess rainwater, stormwater, meltwater, or other sources, can no longer infiltrate into the soil.

Saturation – The state or process that occurs when no more of a substance can be absorbed, combined with, or added.

Sediment – Soil that is moved by erosion and deposited in a new location, often in ditches, rivers, lakes, streams, and other water bodies.

Soil – A dynamic natural medium consisting of mineral particles, air, water, and organic matter and living organisms that is the site of essential regulatory processes in ecosystems.

Soil exposure – Exposure of soils to erosive forces from wind and water.

Soil infiltration – The process by which water on the ground surface enters the soil.

Soil salinization – The process by which water-soluble salts accumulate in the soil.

Soil texture – The relative content of particles of various sizes, such as sand, silt, and clay in the soil.

Solar energy – Radiant energy emitted by the sun or the energy from the sun that strikes the Earth.

Solid – A state of matter that has definite shape and volume.

Soluble – Able to be dissolved, in this text dissolved in water.

Stable – The element or compound resistant to change into another form.

Stream flow – The flow of water in streams, rivers, and other channels. Also the rate or volume of flow.

Stomata – Any of the minute pores in the epidermis of the leaf or stem of a plant, forming a slit of variable width, which allows the movement of gases in and out of the intercellular spaces.

Strata – A series of layers of soil or rock in the ground or Earth's subsurface.

Sub-surface flow – The flow of water beneath the Earth's surface to waterways.

Sugar – Any of the class of sweet-tasting, carbon-containing, water-soluble molecules that comprise carbohydrates, of which sucrose (table sugar) is best known, and which are the primary products of photosynthesis.

Surface ditches – Ditches or channels used to facilitate the drainage of water.

System – A set of things working together as parts of an interconnecting network.

Temperature – The degree or intensity of heat present in a substance or object, especially as expressed according to a comparative scale and shown by a thermometer or perceived by touch.

Terrestrial pools – Reservoirs of various substances (especially carbon, nitrogen, and phosphorus) on the surface or upper layer of Earth generally associated with soil, flora, and fauna.

The Commons – Land or resources not controlled by a single entity but belonging to or affecting a broad population.

Tile drainage – A system of subsurface perferated pipe, laid 3 to 6 feet below the ground surface, that is installed to drain water from cropland into drainage ditches or streams to improve suitability for cropping.

Tillage – The mechanical manipulation of soil for any desired purpose, but generally in agriculture referring to the disturbing of the soil in preparation for planting.

Timescale – The time involved in a process or sequence of events.

Transpiration – The process of water movement through a plant and its evaporation from aerial parts.

Volatilization – The process whereby a dissolved sample transitions from a liquid phase to a vapor phase.

Water cycle – The cycle of processes by which water circulates among the Earth's oceans, atmosphere, soils, and land, involving precipitation as rain and snow, drainage in streams and rivers, and return to the atmosphere by evaporation and transpiration.

Watershed – Area encompassing land upon which rain or snowmelt falls and is transported to creeks, streams, rivers, and eventually to outflow points such as reservoirs, bays, and the ocean.

Weathering – The various mechanical and chemical processes that cause exposed rock to decompose.

Wetland – A distinct ecosystem characterized by flooding by water, either permanently or seasonally, and where low oxygen or anoxic processes prevail.

The definitions are drawn from the sources listed, but numerous definitions have been refined to reflect their specific meanings in environmental management and how they are used in the text.

Dictionary.com (2020). https://www.dictionary.com (accessed 28 August 2020).

NASA (2020). https://www.nasa.gov (accessed 28 August 2020).

Brady (1974). The Nature and Property of Soils, 8th edn. Macmillan Publishing Co. Inc., New York, USA.

Haynes R. J. (2005). Labile Organic Matter Fraction as Central Components of the Quality of Agricultural Soils: An Overview in Advance in Agronomy, Vol. **85**. Donald Sparks (ed.) Elsevier Academic Press, New York, pp. 221–263.

Wikepedia (2020). https://en.wikipedia.org (accessed 28 August 2020).

Rowlings J. K. (1993). *Harry Potter and the Order of Pheonix.*

Chemicool (2020). https://www.chemicool.com (accessed 28 August 2020).

National Geographic (2020). https://www.nationalgeographic.org (accessed 28 August 2020).

Merriam-Webster (2020). https://www.merriam-webster.com (accessed 28 August 2020).

Literature Cited

Abe S. (2001). The First Sulfur Eaters. https://astrobiology.nasa.gov/news/the-first-sulfur-eaters/ (accessed 26 August 2020).

Abramopoulos F., Rosenzweig C. and Choudhury B. (1988). Improved ground hydrology calculations for Global Climate Models (GCMs): soil water movement and evapotranspiration. *Journal of Climate*, **1**(9), 921–941. https://doi.org/10.1175/1520-0442(1988)001<0921:IGHCFG>2.0.CO;2 (accessed 9 August 2020).

Adetunji I. (2020). What Dust from the Sahara Does to You and the Planet, IFLScience. https://www.iflscience.com/environment/what-dust-sahara-does-you-and-planet/ (accessed 16 August 2020).

Agricultural Drainage Management Coalition. (2020). personal communication.

Ahi, ablame L. M., Chaubey I., Smith D. R. and Engel B. A. (2011). Effect of tile effluent on nutrient concentration and retention efficiency in agricultural drainage ditches. *Agricultural Water Management*, **98**(8), 1271–1279. https://www.researchgate.net/publication/227411429_Effect_of_Tile_Effluent_on_Nutrient_Concentration_and_Retention_Efficiency_in_Agricultural_Drainage_Ditches (accessed 12 August 2020).

Aillery M., Gollehon N., Johansson R., Kaplan J., Key N. and Ribaudo M. (2005). Managing Manure to Improve Air and Water Quality, Economic Research Report 9, Economic Research Service, U.S. Department of Agriculture, Washington, D.C., USA. file:///C:/Users/skiph/AppData/Local/Temp/ae030824.pdf (accessed 18 October 2020).

An S., Mentler A., Mayer H. and Blum W. E. H. (2010). Soil aggregation, aggregate stability, organic carbon and nitrogen in different soil aggregate fractions under forest and shrub vegetation on the Loess Plateau, China. *Catena*, **81**(3), 226–233. https://www.sciencedirect.com/science/article/abs/pii/S0341816210000457 (accessed 10 August 2020).

Asbjornsen H., Mora G. and Helmers M. J. (2007). Variation in water uptake dynamics among contrasting agricultural and native plant communities in the Midwestern U.S Agriculture,

© IWA Publishing 2021. A Guide to Understanding the Fundamental Principles of Environmental Management. It Ain't Magic: Everything Goes Somewhere
Authors: Andy Manale and Skip Hyberg
doi: 10.2166/9781789060980_0183

Ecosystems & Environment, **121**(4), 343–356. https://www.sciencedirect.com/science/article/abs/pii/S0167880906004087 (accessed 12 August 2020).

Aulenbach B. T., Buxton H. T., Battaglin W. A. and Coupe R. H. (2007). Streamflow and Nutrient Fluxes of the Mississippi-Atchafalaya River Basin and Subbasins for the Period of Record Through 2005, U.S. Geological Survey Open-File Report 2007-1080, U.S. Geological Survey, U.S. Department of Interior, Washington, D.C., USA. http://toxics.usgs.gov/pubs/of-2007-1080/index.html (accessed 6 July 2020).

Bailey A. W. (1988). Understanding fire ecology for range management. In: Vegetation Science Applications for Rangeland Analysis and Management. Handbook of Vegetation Science. P. T. Tueller (ed.), Springer, Dordrecht, **14**, pp. 527–557. https://link.springer.com/chapter/10.1007/978-94-009-3085-8_21#citeas (accessed 11 August 2020).

Baldwin H. L. and McGuinness C. L. (1963a). A Primer on Ground Water. U.S. Geologic Survey, U.S. Department of Interior, Washington, D.C., USA. https://pubs.usgs.gov/gip/7000056/report.pdf (accessed 9 August 2020).

Baldwin H. L. and McGuinness C. L. (1963b). A Primer on Ground Water. U.S. Geological Survey United States Department of the Interior, Washington, D.C., USA. https://pubs.er.usgs.gov/publication/7000056 (accessed 8 August 2020).

Bales J. D., Oblinger C. J. and Sallenger A. H. (2000). Two Months of Flooding in Eastern North Carolina, September-October 1999 Hydrologic, Water Quality, and Geologic Effects of Hurricanes Dennis, Floyd, and Irene Water, Resources Investigations Report 00-4093, U.S. Geological Survey, U.S. Department of the Interior, Reston, VA, USA.

Balter M. (2015). Researchers Discover First Use of Fertilizer. *Science*. https://www.sciencemag.org/news/2013/07/researchers-discover-first-use-fertilizer (accessed 11 August 2020).

Bauer S. E., Tsugaridis K. and Miller R. (2016). Significant atmospheric aerosol pollution caused by world food production. *AGU Geophysical Research Letters*. **43**(10), 5394–5400. https://doi.org/10.1002/2016GL068354 (accessed 16 August 2020).

BBC Bitesize. (2020). Rivers and Flooding. https://www.google.com/search?q=BBC+Bitesize+water+storm+response+graphs&tbm=isch&ved=2ahUKEwiCzJOQnM3rAhWEAd8KHbneD8gQ2-cCegQIABAA&oq=BBC+Bitesize+water+storm+response+graphs&gs_lcp=CgNpbWcQDFDTyAxY_NQMYOPiDGgAcAB4AIABYIgBmgGSAQEymAEAoAEBqgELZ3dzLXdpei1pbWfAAQE&sclient=img&ei=xQBRX8LVG4SD_Aa5vb_ADA&bih=683&biw=1440&client=firefox-b-1-d#imgrc=RUd4l8zWai6aKM (accessed 3 September 2020).

Belmaker G. (2018). More than Half of Europe's Forests Lost Over 6000 years. https://news.mongabay.com/2018/01/more-than-half-of-europes-forests-lost-over-6000-years/ (accessed 16 August 2020).

Bezdicek D. F. and Kennedy A. C. (1998). Microorganisms in action. In: The Microbial World: The Nitrogen Cycle and Nitrogen Fixation. J. M. Lynch and J. E. Hobbie (eds.), Scientific Publications Produced by Jim Deacon Institute of Cell and Molecular Biology, The University of Edinburgh, Edinburgh, UK. http://archive.bio.ed.ac.uk/jdeacon/microbes/nitrogen.htm (accessed 16 August 2020).

Biello D. (2008). Ocean dead zones continue to spread. *Scientific American*. https://www.scientificamerican.com/article/oceanic-dead-zones-spread/ (accessed 5 June 5 2020).

Birkeland P. W., Shroba R. R., Burns S. F., Price A. B. and Tonkin P. J. (2003). Integrating soils and geomorphology in mountains – an example from the Front Range of Colorado. *Geomorphology*, **55**(1–4), 329–344. https://www.sciencedirect.com/science/article/abs/pii/S0169555X0300148X (accessed August 10).

Blanco-Canqui H., Shaver T. M., Lindquist J. L., Shapiro C. A., Elmore R. W., Francis C. A. and Hergert G. W. (2015). Cover crops and ecosystem services: insights from studies in temperate soils. *Agronomy Journal*, **107**(6), 2449–2474. https://www.researchgate.net/publication/283035604_Cover_Crops_and_Ecosystem_Services_Insights_from_Studies_in_Temperate_Soils (accessed 13 August 2020).

Bond T. (2009). Nitrous Oxide Key Ozone Destroyer. https://www.chemistryworld.com/news/nitrous-oxide-key-ozone-destroyer/3002946.article (accessed 8 August 2020).

Bonne R. 1961. I Know and Old Lady. Rand McNally, Chicago USA.

Boone L. V., Vasilas B. L. and Welch L. F. (1984). The nitrogen content of corn grain as affected by hybrid, population, and location. *Communications in Soil Science and Plant Analysis*, **15**(6), 639–650.

Bot Al. (2005). The Importance of Soil Organic Matter Key to Drought-Resistant Soil and Sustained Food Production, Chapter 5, Creating Drought-Resistant Soil, Soils Bulletin 80, Food and Agriculture Organization, Rome, Italy. http://www.fao.org/3/a0100e/a0100e00.htm#Contents (accessed 3 August 2020).

Boyer J. N. and Groffman P. M. (1996). Bioavailability of water extractable organic carbon fractions in forest and agricultural soil profiles. *Soil Biology and Biochemistry*, **28**(6), 783–790.

Brady N. C. (1974). The Nature and Property of Soils, 8th edn, Macmillion Publishing Co. Inc., New York, USA.

British Petroleum. 2020. BP Statistical review of world energy, www.bp.com/statisticalreview, accessed online on Sept. 3, 2020).

Brooks F. and Jaynes D. (2017). Quantifying the Effectiveness of Installing Saturated Buffers on Conservation Reserve Program to Reduce Nutrient Loading from Tile Drainage Waters, Agricultural Drainage Management Coalition Report AG-3151-P-16-0225t Farm Service Agency, U.S. Department of Agriculture, Washington, D.C., USA. https://www.fsa.usda.gov/Assets/USDA-FSA-Public/usdafiles/EPAS/natural-resouces-analysis/Water/pdfs/2017_Saturated_Buffer_Final_Report.pdf (accessed 13 August 2020).

Buckley K. and Makortoff M. (2004). Phosphorus in Livestock Manures. In: Advanced Silage Corn Management 2004. S. Bittman and C. G. Kowalenko (ed.), Pacific Field Corn Association. Agassiz BC, Canada. https://farmwest.com/node/953 (accessed 18 October 2020).

Bureau of Reclamation. (2020). https://www.usbr.gov/mp/arwec/water-facts-ww-water-sup.html (accessed 3 August 2020).

Burns S. F., Price A. B. and Tonkin P. J. (2003). Integrating soils and geomorphology in mountains – an example from the Front Range of Colorado. *Geomorphology*, **55**(1–4), 329–344. https://www.sciencedirect.com/science/article/abs/pii/S0169555X0300148X (accessed 10 August 2020).

California State Water Resources Control Board. (2011). Runoff Coefficient (C) Fact Sheet Fact Sheet 5.1.3 FS-(RC). https://www.waterboards.ca.gov/water_issues/programs/swamp/docs/cwt/guidance/513.pdf (accessed 20 August 2020).

Canfield D. E., Kristensen E. and Thamdrup B. (2005). The global phosphorus cycle. *Advances in Marine Biology*, **48**, 419–440. https://www.sciencedirect.com/science/article/pii/S0065288105480116#! (accessed 26 August 2020).

Canfield D., Glazer A. N. and Falkowski P. G. (2010). The evolution and future of Earth's nitrogen cycle. *Science*, **330**, 192–196. http://citeseerx.ist.psu.edu/viewdoc/download?doi=10.1.1.465.3125&rep=rep1&type=pdf (accessed 19 August 2020).

Centers for Disease Control and Prevention. (2014). Drinking Water. https://www.cdc.gov/healthywater/drinking/private/wells/ (accessed 3 September 2020).

Cerri C. C., Bernoux M., Arroways D., Feigl B. J. and Piccolo M. C. (2000). Carbon stocks in soils of the Brazilian Amazon. In: Global Climate Change and Tropical Ecosystems. R. Lal, J. M. Kimble and B. A. Stewart (eds.), CRC Press, Washington, D.C., USA, pp. 33–50.

Chemistry Libre Texts. (2020). ATP – the Universal Energy Currency. https://chem.libretexts.org/Bookshelves/Introductory_Chemistry/Book%3A_The_Basics_of_GOB_Chemistry_(Ball_et_al.)/20%3A_Energy_Metabolism/20.02%3A_ATP-_the_Universal_Energy_Currency (accessed 8 August 2020).

Chen X., Dong W., Ou G., Wang Z. and Liu C. (2013). Gaining and losing stream reaches have opposite hydraulic conductivity distribution patterns. *Hydrology Earth and Systems Science*, **17**, 2569–2579. https://chem.libretexts.org/Bookshelves/Introductory_Chemistry/Book%3A_The_Basics_of_GOB_Chemistry_(Ball_et_al.)/20%3A_Energy_Metabolism/20.02%3A_ATP-_the_Universal_Energy_Currency (accessed 8 August 2020).

Christiano D. (2017). Blue Baby Syndrome. https://www.healthline.com/health/blue-baby-syndrome (accessed 21 August 2020).

Christianson L., Helmers M. J., Bhandari A., Kult K., Sutphin T. and Wolf R. (2012). Performance evaluation of four field-scale agricultural drainage denitrification bioreactors in Iowa. *Transactions of the American Society of Agricultural and Biological Engineers*, **55**(6), 2163–2174. https://www.researchgate.net/publication/235721893_Performance_Evaluation_of_Four_Field-Scale_Agricultural_Drainage_Denitrification_Bioreactors_in_Iowa#fullTextFileContent (accessed 13 August 2020).

Clark E. H. (1985). The off-site costs of soil erosion. *Journal of Soil and Water Conservation*, **40**(1), 19–22.

Cleveland C. C. and Liptzin D. (2007). C:N:P stoichiometry in soil: is there a 'Redfield ratio' for the microbial biomass? *Biogeochemistry*, **85**, 235–252. http://citeseerx.ist.psu.edu/viewdoc/download?doi=10.1.1.622.3368&rep=rep1&type=pdf (accessed 13 August 2020).

Concio R. May 17, 2019. The Science Times. https://www.sciencetimes.com/articles/21900/20190517/earths-atmospheric-co2-levels-at-415 ppm-same-as-5-3-million-years-ago.htm (accessed 8 August 8 2020).

Conservation Effects Assessment Project. (2012). Assessment of the Effects of Conservation Practices on Cultivated Cropland in the Upper Mississippi River Basin. Natural Conservation Resources Service. U.S. Department of Agriculture, Washington, D.C., USA. https://www.nrcs.usda.gov/Internet/FSE_DOCUMENTS/stelprdb1042093.pdf (accessed 13 August 2020).

Crumpton W. G., Stenback G. A., Miller B. A. and Helmers M. J. (2006). Potential Benefits of Wetland Filters for Tile Drainage Systems: Impact on Nitrate Loads to Mississippi River

Subbasins. Final project report to U.S. Department of Agriculture Project number: IOW06682. https://www.fsa.usda.gov/Assets/USDA-FSA-Public/usdafiles/EPAS/PDF/fsa_final_report_crumpton_rhd.pdf (accessed 18 October 2020).

Crutzen P. J. (2002). Geology of mankind. *Nature*, **415**, 23. https://www.researchgate.net/publication/11578197_Geology_of_Mankind/references (accessed 19 August 2020).

Dahl T. E. (1990). Wetland Losses in The United States 1780s to 1980s. Fish and Wildlife Service, U.S. Department of the Interior, Washington, D.C., USA. https://www.fws.gov/wetlands/documents/Wetlands-Losses-in-the-United-States-1780s-to-1980s.pdf (accessed 19 August 2020).

Dahl T. E. (2011). Status and Trends of Wetlands in the Conterminous United States 2004 to 2009. Fish and Wildlife Service, U.S. Department of the Interior, Washington, D.C., USA. https://www.fws.gov/wetlands/documents/Status-and-Trends-of-Wetlands-in-the-Conterminous-United-States-2004-to-2009.pdf (accessed 19 August 2020).

Dahl T. E. and Allord G. J. (1997). Technical Aspects of Wetlands History of Wetlands in the Conterminous United States. USGS National Water Summary Paper 2425. Fish and Wildlife Service, U.S. Department of the Interior, Washington, D.C., USA. https://www.fws.gov/wetlands/documents/History-of-Wetlands-in-the-Conterminous-United-States.pdf (accessed 19 August 2020).

Davies D. (2019). Poop and Conquest Citation: The History of American Imperialism, From Bloody Conquest to Bird Poop. National Public Radio. https://www.npr.org/2019/02/18/694700303/the-history-of-american-imperialism-from-bloody-conquest-to-bird-poop (accessed 8 August 2020).

Des Moines Water Works. (2020). Frequently Asked Questions Fact Sheet. http://www.dmww.com/upl/documents/water-quality/lab-reports/fact-sheets/nitrate-removal-facility.pdf (accessed 8 July 2020).

Donigian A. S., Barnwell T. O., Jackson R. B., Patwardhan A. S. and Weinrich K. B. (1994). Assessment of Alternative Management Practices and Policies Affecting Soil Carbon in Agroecosystems of the Central United States. Technical Report. Office of Scientific and Technical Information, Office of Science, U.S. Department of Energy, Washington, D.C., USA. https://www.osti.gov/biblio/7107777 (accessed 16 August 2020).

Dou Z., Galligan D. T., Allshouse R. D., Toth J. D., Ramberg C. F. and Ferguson J. D. (2001). Manure sampling for nutrient analysis: variability and sampling efficacy. *Journal of Environmental Quality*, **30**(4), 1432–1437.

Dunkerley D. L. and Booth T. L. (1999). Plant canopy interception of rainfall and its significance in a banded landscape, arid western New South Wales. *Australia Water Resources Research*, **35**(5), 1581–1586. https://agupubs.onlinelibrary.wiley.com/doi/pdf/10.1029/1999WR900003 (accessed 11 August 2020).

Ecclesiastes 1:7, King James version of the Bible (1611).

Edwards P. J., Willard K. W. J. and Schoonover J. E. (2015). Fundamentals of watershed hydrology. *Journal of Contemporary Water Research and Education*, **154**, 3–20.

Eni School. (2020). Soil Formation. http://www.eniscuola.net/en/argomento/soil/soil-formation/how-long-does-it-take-to-form/ (accessed 20 August 2020).

European Union. (2007). Soil Protection – a New Policy for the EU. https://ec.europa.eu/environment/soil/pdf/factsheet_2007_en.pdf (accessed 20 August 2020).

Fang Z., Li D., Jiao F., Yao J. and Du H. (2019). The latitudinal patterns of leaf and soil C:N:P stoichiometry in the Loess Plateau of China. *Frontiers of Plant Science*, **10**, 85. https://doi.org/10.3389/fpls.2019.00085 (accessed 9 August 2020).

Federal Reserve Bank of St. Louis. (2020). Economic Research: Resident Population of Wake County, NC. https://fred.stlouisfed.org/series/NCWAKE3POP (accessed 31 August 2020).

Field C., Behrenfeld M. J., Randerson J. T. and Falkowski P. (1998). Integrating terrestrial and oceanic components. *Science*, **10**(281), 237–240. https://science.sciencemag.org/content/281/5374/237 (accessed 31 August 2020).

Fields S. (2004). Global nitrogen: cycling out of control. *Environmental Health Perspectives*, **112**(10) A556–A563. https://www.ncbi.nlm.nih.gov/pmc/articles/PMC1247398/ (accessed 31 August 2020).

Fischer D. and Hutjens M. (2019). How Many Pounds of Feed Does a Cow Eat in a Day? University of Illinois – Illini DairyNET https://dairy-cattle.extension.org/how-many-pounds-of-feed-does-a-cow-eat-in-a-day/#:~:text=As%20cows%20produce%20more%20milk%2C%20they%20eat%20more.,Educator%2C%20University%20of%20Illinois%20%E2%80%93%20Illini%20DairyNET%20and (accessed 31 August 2020).

Food and Agriculture Organization. (2003). Crop Production and Natural Resource Use. In: World Agriculture: Toward 2015/2030 – An FAO perspective World Agriculture. J. Bruinsma (ed.), Earthscan Publications Ltd, London. http://www.fao.org/3/y4252e/y4252e06.htm (accessed 3 August 2020).

Food and Agriculture Organization. (2017). World Fertilizer Trends and Outlook to 2020: Summary Report. United Nations, FAO, Rome, Italy. http://www.fao.org/publications/card/en/c/cfa19fbc-0008-466b-8cc6-0db6c6686f78/ (accessed 8 August 2020).

Foreign Agricultural Service. (2020). Market and Trade Data. https://apps.fas.usda.gov/psdonline/app/index.html#/app/compositeViz (accessed 20 August 2020).

Fowler D., Mhairi C., Ute S., Sutton M. A., Cape J. N., Reis S., Sheppard L. J., Jenkins A., Grizzetti B., Galloway J. N., Vitousek P., Leach A., Bouwman A. F., Butterback-Bahi K., Dentener F., Stevenson D., Amann M. and Voss M. (2013). The global nitrogen cycle in the twenty-first century. *Philosophical Transactions of the Royal Society B*, **368**, 20130164. http://doi.org/10.1098/rstb.2013.0164 (accessed 8 August 2020).

Franche C., Lindström K. and Elmerich C. (2009). Nitrogen-fixing bacteria associated with leguminous and non-leguminous plants, *Plant Soil*, **321**, 35–59.

Gavrilescu M. (2014). Colloid-mediated transport and the fate of contaminants in soils. In: The Role of Colloidal Systems in Environmental Protection. M. Fanun (ed.), Elsevier, Amsterdam, Netherlands, pp. 397–451.

Gende S. M., Edwards R. T., Willson M. F. and Wipfli M. S. (2002). Pacific Salmon in aquatic and terrestrial ecosystems: Pacific Salmon subsidize freshwater and terrestrial ecosystems through several pathways, which generates unique management and conservation issues but also provides valuable research opportunities. *BioScience*, **52**(10), 917–928. https://doi.org/10.1641/0006-3568(2002)052[0917:PSIAAT]2.0.CO;2 (accessed 8 August 2020).

Gleich P. (2017). Water resources, climate change, and the destabilization of Modern Mesopotamia. In: Water, Security and U.S. Foreign Policy. D. Reed (ed.), Routledge, Vol. **1**, pp. 149–167. https://warroom.armywarcollege.edu/articles/water-u-s-national-security/ (accessed 22 August 2020).

Goldberg, S.R., Lebron, I., Seaman, J.C., Suarez, D.L. 2012. Soil colloidal behavior. In: P.M. Huang, Y. Li and M.E. Sumner (eds.) Handbook of Soil Sciences Properties and Processes (2nd Edition). CRC Press, Taylor and Francis Group. Boca Raton, FL. Chapter 15. pp. 15-1–15-39.

International Geosphere-Biosphere Program. (2020). Global Change: Anthropocene. http://www.igbp.net/globalchange/anthropocene.4.1b8ae20512db692f2a680009238. html (accessed 19 August 2020).

Greenan C. M., Moorman T. B., Parkin T. B., Kaspar T. C. and Jaynes D. B. (2009). Denitrification in wood chip bioreactors at different water flows. *Journal of Environmental Quality*, **38**, 1664–1671. https://www.researchgate.net/publication/ 26314399_Denitrification_in_Wood_Chip_Bioreactors_at_Different_Water_Flows (accessed 13 August 2020).

Griffiths B. S., Spilles A. and Bonkowski M. (2012). C:N:P stoichiometry and nutrient limitation of the soil microbial biomass in a grazed grassland site under experimental P limitation or excess. *Ecological Processes*, **1**, 6. https://doi.org/10.1186/2192-1709-1-6, https://ecologicalprocesses.springeropen.com/articles/10.1186/2192-1709-1-6 (accessed 9 August 2020).

Gruber N. and Deutsch C. A. (2014). Redfield's evolving legacy. *Nature Geoscience*, **7**, 853–855. https://doi.org/10.1038/ngeo2308 (accessed 9 August 2020).

Hansen L. and Hellerstein D. (2007). The value of the reservoir services gained with soil conservation. *Land Economics*, **83**(3), 285–301.

Hansen L., Breneman V. E., Davison C. W. and Dicken C. W. (2002). The cost of soil erosion to downstream navigation. *Journal of Soil and Water Conservation*, **57**(4), 205–212.

Harari Y. N. (2015). Sapiens: A Brief History of Humankind. Harper Perennial, New York, USA.

Hardin G. (1868). "The Tragedy of the Commons". *Science*, **62**(3859), 1243–1248.

Hatfield J. L., Jaynes D. B., Burkart M. R., Cambardella C. A., Moorman T. B., Prueger J. H. and Smith M. A. (1999). Water Quality in Walnut Creek Watershed: Setting and Farming Practices. *Journal of Environmental Quality*, **28**(1) 11–28.

Hatfield J., McMullen L. D. and Jones C. S. (2009). Nitrate-nitrogen patterns in the Raccoon River Basin related to agricultural practice. *Journal of Soil and Water Conservation*, **64**(3), 190–199.

Haynes R. J. (2005). Labile organic matter fraction as central components of the quality of agricultural soils: an overview. In: Advances in Agronomy. D. Sparks (ed.), Elsevier Academic Press, New York, USA, Vol. **85**, pp. 221–268.

Hayward K. (2013). Narrow Tube Openings Allow Capillary Action to Pull Water Higher. https://www.usgs.gov/media/images/narrower-tube-openings-allow-capillary-action-pull-water-higher (accessed 3 September 2020).

Heath L. S., Kimble J. M., Birdsey R. A. and Lal R. (2003). The Potential of U.S. Forest Soils to Sequester Carbon. In: The Potential of US. Forest Soils to Sequester Carbon and Mitigate the Greenhouse Effect. J. M. Kimble, L. S. Heath, R. A. Birdsey and R. Lal (eds.), CRC Press. Washington DC, USA pp. 385–394. https://www.fs.fed. us/ne/newtown_square/publications/other_publishers/OCR/ne_2003heath01.pdf (accessed 18 October 2020).

Henderson L. (1913). The Fitness of the Environment, an Inquiry into the biological significance of the properties of matter. Macmillan, New York, USA.

Heimlich R. E. and Daugherty A. B. (1991). America's cropland: where does it come from? In: Agriculture and the Environment: Yearbook of Agriculture 1991. D. T. Smith (ed.), U.S. Department of Agriculture, Washington, D.C., USA, pp. 3–9. https://naldc.nal. usda.gov/download/IND20394000/PDF (accessed 11 August 2020).

Hoffman G. J., Howell T. A. and Solomon K. H. (1990). Introduction. In: Management of Farm Irrigation Systems. G. J. Hoffman, T. A. Howell and K. H. Solomon (eds.), American Society of Agricultural Engineers, St. Joseph, MI, USA, pp. 5–10.

Hollis G. (1988). Rain, roads, roofs and runoff: hydrology in cities. *Geography*, **73**(1), 9–18. www.jstor.org/stable/40571337 (accessed 12 August 2020).

Huszar P. C. and Piper S. L. (1986). Estimating the off-site costs of wind erosion in New Mexico. *Journal of Soil and Water Conservation*, **41**(6), 414–416. https://www. jswconline.org/content/41/6/414.short (accessed 13 August 2020).

Hyberg S., Iovanna R., Crumpton W. and Richmond S. 2015. The cost effectiveness of wetlands designed and sited for nitrate removal: The effect of increased efficiency, rising easement costs, and lower interest rates. *Journal of Soil and Water Conservation*, **70**(2), 30A–32A. https://www.jswconline.org/content/70/2/30A (assessed 17 October 2020).

Immerwahr D. (2020). How to Hide and Empire. Farrar Straus and Giroux – Macmillan Publishers, New York, USA.

Interagency Floodplain Management Review Committee. (1994). Sharing the Challenge: Floodplain Management into the 21st Century, Report to the Administration Floodplain Management Task Force, Washington, D.C., USA. https://fas. org/irp/agency/dhs/fema/sharing.pdf (accessed 13 August 2020).

Jackson R., Saunois M., Bousquet P., Canadell P. and Poulter B. (July 15, 2020). Methane Is on an Alarming Upward Trend. Scientific American. https://www.scientificamerican. com/article/methane-is-on-an-alarming-upward-trend/ (accessed 3 September 2020).

Jarvie H. P., Johnson L. T., Sharpley A. N., Smith D. R., Baker D. B., Bruulsema T. W. and Confesor R. (2017). Increased soluble phosphorus loads to Lake Erie: unintended consequences of conservation practices? *Journal of Environmental Quality*, **46(1)**, **123–132**. https://acsess.onlinelibrary.wiley.com/doi/full/10.2134/jeq2016.07.0248 (accessed 12 Jul 2020).

Jayasinghe S., Miller D. and Hatfield J. L. (2012). Evaluation of variation in nitrate concentration levels in the Raccoon River Watershed in Iowa. *Journal of Environmental Quality*, **41**, 1557–1565.

Jones R., Frarey L., Bouzaher A., Johnson S. and Neibergs S. (1993). Livestock and The Environment: A National Pilot Project Detailed Problem Statement: Project Task 1.1 Texas Institute for Applied Environmental Research Tarleton State University Box T0410, Tarleton Station Stephenville, and Center for Agricultural and Rural Development Iowa State University, Ames, IA, USA. http://tiaer.tarleton. edu/pdf/PR9304.pdf (accessed 16 July 2020).

Kalamees R., Püssa K., Zobel K. and Zobel M. (2012). Restoration potential of the persistent soil seed bank in successional calcareous (alvar) grasslands in Estonia. *Applied Vegetation Science*, **15**(2), 208–218. https://www.researchgate.net/publication/ 257935498_Restoration_potential_of_the_persistent_soil_seed_bank_in_successional_ calcareous_alvar_grasslands_in_Estonia (accessed 10 August 2020).

Kansas State University Soil Carbon Center. (2020). What is the Carbon Cycle? https:// soilcarboncenter.k-state.edu/carbcycle.html (accessed 8 August 2020).

Kellogg R. L., Lander C. H., Moffitt D. C. and Gollehon N. (2000). Manure nutrients relative to the capacity of cropland and pastureland to assimilate nutrients: spatial and temporal trends for the United States. *Proceedings of the Water Environment Federation*, **16**, 18–157. https://www.researchgate.net/publication/233583223_Manure_Nutrients_Relative_to_The_Capacity_Of_Cropland_And_Pastureland_To_Assimilate_Nutrients_Spatial_and_Temporal_Trends_for_the_United_States (accessed 12 August 2020).

Kim J., Johnson L., Cifelli R., Thorstensen A. and Chandrasekar V. (2019). Assessment of antecedent moisture condition on flood frequency: an experimental study in Napa River Basin, CA. *Journal of Hydrology*, **26**, 1–18. https://www.sciencedirect.com/science/article/pii/S2214581818303768#! (accessed 10 August 2020).

Kirchner J. W. and Allen S. T. (2020). Seasonal partitioning of precipitation between streamflow and evapotranspiration, inferred from end-member splitting analysis. *Hydrological Earth System Science*, **24**, 17–39. https://hess.copernicus.org/articles/24/17/2020/ (accessed 21 August 2020).

Kleinman P. J., Brooks R., Mccarty G. W., Veith T. L., Fernandez C., Hagan E., Nassry M., Saporito L. S., Wallace C., Iovanna R. and Hyberg S. (2018). Effectiveness of conservation reserve program buffers in the Chesapeake Bay Watershed: 2017 Annual Report, Government Publication/Report. P.1, U.S. Agricultural Research Service, U.S. Department of Agriculture, Washington, D.C., USA. https://www.ars.usda.gov/research/publications/publication/?seqNo115=353802 (accessed 22 August 2020).

Knopf F. L. and Sampson F. B. (1996). Conservation of grassland vertibrates. In: Prairie Conservation: Preserving North America's Most Endangered Ecosystem. R. Samson and F. L. Knopf (eds.), Island Press, Washington, D.C., USA, pp. 273–289. https://www.fws.gov/southwest/es/documents/R2ES/LitCited/LPC_2012/Knopf_and_Samson_1997.pdf (accessed 23 August 2020).

Koelsch R. (2018). What is the Economic Value of Cattle Manure? Drovers.com. Milk Business Conference. https://www.drovers.com/article/what-economic-value-cattle-manure (accessed 18 October 2020).

Kramer J. and Weaver J. E. (1936). Relative efficiency of roots and tops of plants in protecting the soil from erosion. Bulletin 12 Conservation Department of the Conservation and Survey Division. University of Nebraska. Lincoln NE, USA. https://digitalcommons.unl.edu/cgi/viewcontent.cgi?article=1003&context=agronweaver (accessed 18 October 2020).

Kramer J. and Weaver J. (2015). Can plants help slow soil erosion? *Scientific American*. https://www.scientificamerican.com/article/can-plants-help-slow-soil-erosion/ (accessed 10 August 2020).

Krissansen-Totten J., Arney G. N. and Catling D. C. (2018). Constraining the climate and ocean pH of the early Earth with a geological carbon cycle model. Proceedings of the National Academy of Science, **115**(16), 4105–4110 https://pubmed.ncbi.nlm.nih.gov/29610313/ (accessed 4 September 2020).

Lal R., Kimble J. M. and Follett R. F. (1997). Land use and soil c pools in terrestrial ecosystems. In: Management of Carbon Sequestration in Soil. R. Lal, J. M. Kimble, R. F. Follett and B. A. Stewart (eds.), Advances in Soil Science Series from CRC Press, New York, USA, pp. 1–10.

Lal R., Kimble J. M., Follett R. F. and Cole C. V. (1998a). The Potential of US Cropland to Sequester Carbon and Mitigate the Greenhouse Effect. Sleeping Bear Press (Ann Arbor Press), Chelsea, MI, USA.

Lal R., Smith P., Jungkunst H. F., Mitsch W., Lehmann J., Nair P. K. R., McBratney A. B., Sá J. C. M., Schneider J., Zinn Y. L., Skorupa A. L. A., Zhang H., Minasny B., Srinivasrao C. and Ravindranath N. H. (1998b). The carbon sequestration potential of terrestrial ecosystems. *Journal of Soil and Water Conservation*, **73**(6), 145–152. https://www.researchgate.net/publication/328809135_The_carbon_sequestration_potential_of_terrestrial_ecosystems (accessed 16 August 2020).

Larson L. W. (1996). 'The Great USA Flood of 1993'. Presented at IAHS Conference Destructive Water: Water-Caused Natural Disasters – Their Abatement and Control, National Weather Service, Anaheim, CA, USA, 24–28 June 1996. https://www.nwrfc.noaa.gov/floods/papers/oh_2/great.htm (accessed 12 August 2020).

Letson D. and Gollehon N. (2007). Spatial economics of targeting manure policy. *Journal of the American Water Resources Association*, **34**(1), 185–193. https://www.researchgate.net/publication/229557625_Spatial_economics_of_targeting_manure_policy (accessed 12 August 2020).

Li X., Xiao Q., Niu J., Dymond S., McPherson E. G., Doorn N., Yu X., Xie B., Zhang K. and Li J. (2017). Rainfall interception by tree crown and leaf litter: an interactive process. *Hydrological Processes*, **31**, 3533–3542. https://www.fs.fed.us/psw/publications/dymond/psw_2017_dymond002_li.pdf (accessed 11 August 2020).

Lilliston B. (2019). Latest Agriculture Emissions Data Show Rise of Factory Farms. https://www.iatp.org/blog/201904/latest-agriculture-emissions-data-show-rise-factory-farms (accessed 16 July 2020).

Lippsett L. (2012). Storms, floods, and droughts: the cycle that transports water around the Earth is intensifying, *OceanUS Magazine*, **49**(3), 20–25. https://www.whoi.edu/oceanus/feature/storms-floods-and-droughts/ (accessed 10 August 2020).

Liu Y. and Chen J. (2008). Phosphorus cycle. In: Encyclopedia of Ecology. M. Schaechter (ed.), Elsevier, Inc., Amsterdam, Netherlands, pp. 2715–2723. https://books.google.com/books?hl=en&lr=&id=6IQY8Uh1aA0C&oi=fnd&pg=PR1&dq=sciencedirect+encyclopedia+of+ecology&ots=sIcCjfak9s&sig=ZI3Zw7hrKJkIkWRpaPoUcWxThyU#v=onepage&q=sciencedirect%20encyclopedia%20of%20ecology&f=false (accessed 8 April 2020).

Lorimar J., Powers W. and Sutton A. (2004). MidWest Plan Service. Iowa State University, Ames, IA, USA. https://www-mwps.sws.iastate.edu/catalog/manure-management/manure-characteristics-pdf (accessed 9 August 2020).

Mackey K. R. M. and Paytan A. (2009). Phosphorus cycle. In: Encyclopedia of Microbiology, Phosphorus Sources, Sinks, and Transport Pathways. M. Schaechter (ed.), 3rd edn, Elsevier, Inc., Amsterdam, Netherlands, pp. 322–334. https://www.sciencedirect.com/topics/agricultural-and-biological-sciences/phosphorus-cycle (accessed 9 August 2020).

Manale A (2010). Fixing Chesapeake Bay: Getting the Strategy Right. In: RFF: Resources 12.26.10. https://www.resourcesmag.org/common-resources/fixing-chesapeake-bay-getting-the-strategy-right/ (accessed online 10 October 2020).

Manale A, Morgan C, Sheriff G, and Simpson D. (2011). Offset markets for nutrient and sediment dischargesin the Chesapeake Bay Watershed: Policy tradeoffs and potential steps forward. National Center For Environmental Economics Working Paper Series, Working Paper # 11-05. https://www.epa.gov/sites/production/files/2014-12/documents/offset_markets_for_nutrient_and_sediment_discharges.pdf (accessed online 10 October 2020.

Mann M. E. (2019). Greenhouse Gas. https://www.britannica.com/science/greenhouse-gas (accessed 9 August 2020).

Martínez-Granados D., Maestre-Valero J. F., Calatrava J. and Martínez-Alvarez V. (2011). The economic impact of water evaporation losses from water reservoirs in the Segura Basin, SE Spain. *Water Resource Management*, **25**(13), 3153–3175. https://doi.org/10.1007/s11269-011-9850-x (accessed 12 August 2020).

McDermott-Murphy C. (2019). No Laughing Matter. https://chemistry.harvard.edu/news/trouble-thaw (accessed 22 August 2020).

Milchunas D. G., Vandever M., Ball L. and Hyberg S. (2011). Allelopathic Cover Crop Prior to Seeding is More Important Than Subsequent Grazing/Mowing in Grassland Establishment. Rangeland Ecology & Management, **64**(3), 291–300. https://www.researchgate.net/publication/228461907_Allelopathic_Cover_Crop_Prior_to_Seeding_Is_More_Important_Than_Subsequent_GrazingMowing_in_Grassland_Establishment (accessed 18 October 2020).

Millennium Ecosystem Assessment – World Health Organization. (2003). Ecosystem and Human Well-Being: A Framework for Assessment. Island Press, Washington, D.C., USA.

Miller D. A. and White R. A. (1998). A conterminous United States multilayer soil characteristics dataset for regional climate and hydrology modeling. *Earth Interact*, **2**(2), 1–26. https://doi.org/10.1175/1087-3562(1998)002<0001:ACUSMS>2.3.CO; (accessed 9 August 2020).

Mitsch W. J., Day J. W., Zhang L. and Lane R. R. (1999). Nitrate-nitrogen retention in wetlands in the Mississippi River Basin. Ecological Engineering, **24**, 267–278. https://d1wqtxts1xzle7.cloudfront.net/39786412/Nitrate-Nitrogen_Retention_in_Wetlands_i20151107-10139-cigyvb.pdf?1446942478=&response-content-disposition=inline%3B+filename%3DNitrate_nitrogen_retention_in_wetlands_i.pdf&Expires=1603037250&Signature=d71ZWAIULlkCfY9eHH3urs~pdBaTJAKbLmf6OIUdlXG3M913a7vJhK9fFO~M1ozBQdGgDxIY~hsxjH~ngVBgpRegNxrv1xkYgyyVar4ZN-Fkpuf7JxpC-n6k0JRQPtFI0bH7yQ2xxsn2v1c1vatj8W9TmxSf~2eb9LWPNFVHbdD2VMwfRRTAmLanXUmhojMFIAP4-4v0ODXnmQGVhjzTq15e6hRW61FXy49k512az1ceoG6 V-vmR~IcoiUBkMzfXcFd6y0M4bnsXocHo~ZeDDa29jFY78hmsJLLwzGLjWSlMMeRrkC5YOsXO7XLR5skW8PRFG4CD0eLG6mHgyEQbGQ__&Key-Pair-Id=APKAJLOHF5GGSLRBV4ZA (accessed 18 October 2020).

Mo, hamadi M. A. and Kavian A. (2015). Effects of rainfall patterns on runoff and soil erosion in field plots. *International Soil and Water Conservation Research*, **3**(4), 273–281. https://www.sciencedirect.com/science/article/pii/S209563391530071X (accessed 12 August 2020).

Mukhtar S. and Auvermann B. W. (2020). Improving the Air Quality of Animal Feeding Operations with Proper Facility and Manure Management. https://agrilifeextension.tamu.edu/library/ranching/improving-the-air-quality-of-animal-feeding-operations-with-proper-facility-and-manure-management/ (accessed 30 July 2020).

Nafziger E. (2017). New Grain Phosphorus and Potassium Numbers. The Bulletin. Farmdoc. University of Illinois. Champaign IL, USA. https://farmdoc.illinois.edu/field-crop-production/uncategorized/new-grain-phosphorus-and-potassium-numbers.html (accessed 20 August 2020).

Nahlik A. and Fennessy M. (2016). Carbon storage in US wetlands. *Nature Communications* **7**, 13835. https://doi.org/10.1038/ncomms13835 (accessed 9 August 2020).

National Agricultural Statistics Service. (2019a). 2017 Census of Agriculture: United States Summary and State Data. Volume **1**, Geographic Area Series, Part 51. AC-17-A 51, National Agricultural Statistics Service, U.S. Department of Agriculture, Washington, D.C., USA. www.agcensus.usda.gov – https://www.nass.usda.gov/Publications/ AgCensus/2017/Full_Report/Volume_1,_Chapter_1_US/st99_1_0047_0047.pdf#page =1&zoom=auto,-150,792 (accessed 9 August 2020).

National Agricultural Statistics Service. (2019b) Agricultural Census: Texas County Profiles. https://www.nass.usda.gov/Publications/AgCensus/2012/Online_Resources/County_ Profiles/Texas/cp48143.pdf (accessed 9 August 2020).

National Agricultural Statistics Service. (2020) Texas Sorghum Yield and Production. https://www.nass.usda.gov/Statistics_by_State/Texas/Publications/Charts_&_Maps/ zsorg_yp.php (accessed 23 August 2020).

National Geographic. (2020). The Ecological Benefits of Fire. https://www. nationalgeographic.org/article/ecological-benefits-fire/ (accessed 25 August 2020).

National Geographic Society Resource Library Encyclopedic. (2020). Petroleum. https:// www.nationalgeographic.org/encyclopedia/petroleum/ (accessed 8 August 2020).

National Oceanic and Atmospheric Administration. (2000). Atmospheric Ammonia: Sources and Fate: A review of Ongoing Federal Research and Future Needs. Air Quality Research Subcommittee Meeting Report, Committee on the Environment and Natural Resources, Air Quality Research Subcommittee, National Oceanic and Atmospheric Administration, U.S. Department of Commerce, Boulder, CO, USA. https://www.esrl.noaa.gov/csl/aqrsd/reports/ammonia.pdf (accessed 19 August 2020).

National Oceanic and Atmospheric Administration. (2002) National Centers for Environmental Information. https://www.ncdc.noaa.gov/news/drought-monitoring-economic-environmental-and-social-impacts (accessed 1 June 2020).

National Oceanic and Atmospheric Administration, Global Monitoring Laboratory. (2020). NOAA's Annual Greenhouse Gas Index. http://www.esrl.noaa.gov/gmd/aggi/aggi. fig2.png (accessed 22 August 2020).

National Oceanic and Atmospheric Administration. (2020). National Centers for Environmental Information. https://www.ncdc.noaa.gov/billions/summary-stats (accessed 11 August 2020).

National Park Service. (2020a). Badlands Mixed-Grass Prairies. https://www.nps.gov/ articles/000/prairie-ecology-badl.htm (accessed 23 August 2020).

National Park Service. (2020b). Tallgrass Prairie: Last Stand of the Tallgrass Prairie. https:// www.nps.gov/tapr/index.htm (accessed 23 August 2020).

National Soil Survey Center. (1995). Soil Survey Geographic (SSURGO) Data Base Data Use Information, Miscellaneous Publication 1527, National Soil Survey Center, Natural Resources Conservation Service, U.S. Department of Agriculture, Washington, D.C., USA. http://www.fsl.orst.edu/pnwerc/wrb/metadata/soils/ssurgo.pdf (accessed 18 August 2020).

National Space and Atmospheric Administration, Earth Observatory. (2005) High Water: Building a Global Flood Atlas. https://earthobservatory.nasa.gov/features/HighWater (accessed 21 August 2020).

National Space and Atmospheric Administration. (2008). Aquatic Dead Zones. https:// earthobservatory.nasa.gov/images/44677/aquatic-dead-zones (accessed 21 August 2020).

National Space and Atmospheric Administration. (2019). The Ocean's Carbon Balance. www.earthobservatory.nasa.gov (accessed 2 December 2019).

National Space and Atmospheric Administration. (2020) Global Climate Change, https:// climate.nasa.gov/evidence/ (accessed 18 October 2020).

Natural Resources Conservation Service. (2013). Soil Health Matters: Add Organic Matter. https://www.nrcs.usda.gov/Internet/FSE_DOCUMENTS/stelprdb1083169.pdf (accessed 5 June 2020).

Natural Resources Conservation Service. (2018). Summary Report: 2015 National Resources Inventory, Natural Resources Conservation Service, U U.S. Department of Agriculture and Center for Survey Statistics and Methodology, Iowa State University, Ames, IA, USA. https://www.nrcs.usda.gov/Internet/FSE_DOCUMENTS/nrcseprd1422028.pdf (accessed 3 August 2020).

Natural Resources Conservation Service. (2020a). Conservation Compliance for Highly Erodible Lands. https://www.nrcs.usda.gov/wps/portal/nrcs/detail/national/programs/ farmbill/?cid=nrcseprd1542214 (accessed 18 August 2020).

Natural Resources Conservation Service. (2020b). Nutrient Management: Natural Resources Conservation Service Conservation Practice Standard. https://www.nrcs. usda.gov/Internet/FSE_DOCUMENTS/stelprdb1192371.pdf (accessed 22 August 2020).

Natural Resources Conservation Service. (2020c). Soil Organic Matter. https://www.nrcs. usda.gov/Internet/FSE_DOCUMENTS/nrcs142p2_053264.pdf (accessed 31 August 2020).

Neal A. L., Bacq-Labreuil A., Zhang X., Clark I. M., Coleman K., Mooney S. J., Ritz K. and Crawford J. W. (2020). Soil as an extended composite phenotype of the microbial metagenome. *Scientific Reports*, **10**, 10649. https://www.nature.com/articles/s41598-020-67631-0 (accessed 8 August 2020).

Neilsen R. 2005. Can we Feed the World? The nitrogen limit of food production', http:// home.iprimus.com.au/nielsens/ (accessed 17 October 2020).

Neuse River Soil and Water Conservation District, Soil Conservation Service U.S. Department of Agriculture, U.S. Forest Service, U.S. Department of Agriculture, County Commissioners of Wake City and of Raleigh, and Crabtree Creek Watershed Improvement District. (1964). Watershed Work Plan Crabtree Creek Watershed Durham and Wake Counties, North Carolina. https://archive.org/stream/CAT91 949408/CAT91949408_djvu.txt (accessed 31 August 2020).

O'Geen A. T. (2013). The Nature Education Knowledge Project, soil water dynamics. *Nature Education Knowledge*, **4**(5), 9. http://www.Nature.com (accessed 31 March 2020).

Oros B. (2020). Pork FAQs. http://www.boboros.com/meatfaqs/pork-017-food-conversion. htm (accessed 20 August 2020).

Oswald S. N. and Smith W. B. (2014). U.S. Forest Resource Facts and Historical Trends. United States Department of Forest Service FS-1035, U.S. Department of Agriculture, Washington, D.C., USA. https://www.fia.fs.fed.us/library/brochures/docs/2012/Forest Facts_1952-2012_English.pdf (accessed 9 August 2020).

Padgitt M. (1997). Soil management and conservation. In: Agricultural Resources and Environmental Indicators. M. Anderson and R. Magleby (eds.), Agricultural Handbook No. AH-712, Economic Research Service, U.S. Department of Agriculture, Washington, D.C., USA, pp. 1–54. https://www.ers.usda.gov/webdocs/publications/41882/30078_arei4-2.pdf?v=0 (accessed 11 August 2020).

Panuska J. (2017). The Basics of Agricultural Tile Drainage: Basic Engineering Principals 2. University of Wisconsin Extension, Madison, WI, USA. https://fyi.extension.wisc.edu/drainage/files/2015/09/Basic_Eng_-Princ-2_2017.pdf (accessed 20 August 2020).

Pennisi E. (2020). Humans Have Altered North America's Ecosystems More Than Melting Glaciers. https://www.sciencemag.org/news/2020/08/humans-have-altered-north-america-s-ecosystems-more-melting-glaciers (accessed 19 August 2020).

Perkins C. 2018. Hay meadows require higher fertility level than pastures, as nutrients are removed with the forage. Tyler Morning Telegraph and Texas A&M AgriLife Extension Service, https://tylerpaper.com/news/local/hay-meadows-require-higher-fertility-level-than-pastures-as-nutrients-are-removed-with-the-forage/article_79c655ce-2168-11e8-8095-9fd4fce19ad7.html (accessed 1 November 2020).

Pimentel D., Harvey C., Resosudarmo P., Sinclair K., Kurz D., McNair M., Crist S., Shpritz L., Fitton L., Saffouri R. and Blair R. (1995). Environmental and economic costs of soil erosion and conservation benefits. *Science*, **267**(5201), 1117–1123. https://www.researchgate.net/publication/6051747_Environmental_and_Economic_Cost_of_Soil_Erosion_and_Conservation_Benefits (accessed 19 August 2020).

Piper S. (1989). Measuring particulate pollution damage from wind erosion in the western United States. *Journal of Soil and Water Conservation*, **44**(1), 70–75. https://www.jswconline.org/content/44/1/70.short (accessed 13 August 2020).

Potter K. N., Torbert H. A., Johnson H. B. and Tischler C. R. (1999). Carbon storage after long-term grass establishment on degraded soils. *Soil Science*, **164**(10), 718–725. https://journals.lww.com/soilsci/Abstract/1999/10000/CARBON_STORAGE_AFTER_LONG_TERM_GRASS_ESTABLISHMENT.2.aspx (accessed 10 August 2020).

Rafferty J. (2020). Ocean Acidification. https://www.britannica.com/science/ocean-acidification (accessed 8 August 2020).

Raleigh News and Observer. (2006). Creek's Rise Is Not a Shock. http://www.newsobserver.com/2006/06/18/42475/creeks-rise-is-not-a-shock.html (accessed 8 August 2020).

Rayner P., Trudinger C., Etheridge D. and Rubino M. (2016). Land Carbon Storage Swelled in the Little Ice Age, Which Bodes Ill for the Future. https://theconversation.com/land-carbon-storage-swelled-in-the-little-ice-age-which-bodes-ill-for-the-future-62965 (accessed 16 August 2020).

Reddy K. R., Newman S., Osborne T. Z., Whitc J. R. and Fitz H. C. Phosphorous Cycling in the Greater Everglades Ecosystem: Legacy Phosphorous Implications for Management and Restoration. (1999). *Critical Reviews in Environmental Science and Technology* **41**(6) 149–186. https://soils.ifas.ufl.edu/wetlands/publications/PDF-articles/356.Phosphorus%20cycling%20in%20the%20Everglades%20ecosystem.pdf (accessed 18 October 2020).

Ribaudo M. O. (1986). Reducing Soil Erosion: Offsite Benefits. Agricultural Economic Report No. 561. Economic Research Service, U.S. Department of Agriculture, Washington, D.C., USA.

Ribaudo M. O., Gollehon N. R. and Agapoff J. (2003a). Land application of manure by animal feeding operations: is more land needed? *Journal of Soil and Water Conservation*, **58**(1), 30–38. https://www.researchgate.net/publication/43289774_Land_application_of_manure_by_animal_feeding_operations_Is_more_land_needed (accessed 12 August 2020).

Ribaudo M., Kaplan J. D., Christensen L. A., Gollehon N., Johansson R., Breneman V., Aillery M., Agapoff J. and Peters M. (2003b). Manure Management for Water Quality Costs to Animal Feeding Operations of Applying Manure Nutrients to Land. Agricultural Economic Report No. 824. Economic Research Service, USDA, Washington, D.C., USA. https://papers.ssrn.com/sol3/papers.cfm?abstract_id=757884 (accessed 12 August 2020).

Rice C., Staffenborg S. and Watson S. (2009) Soils Emitting More Carbon Dioxide. https://soilcarboncenter.k-state.edu/originals/Soils_emitting_more_CO2.html (accessed 25 August 2020).

Ritchie H. 2017. Fossil Fuels. Our World in Data, https://ourworldindata.org/fossil-fuels, (accessed 2 September 2020).

Roberts N., Fyfe R. M., Woodbridge J., Gaillard M. J., Davis B. A. S., Kaplan J. O., Marquer L., Mazier F., Nielsen A. B., Sugita S., Trondman A. K. and Leydet M. (2018). Europe's lost forests: a pollen-based synthesis for the last 11,000 years. *Science Reports*, **8**, 716. https://doi.org/10.1038/s41598-017-18646-7 (accessed 9 August 2020).

Robertson G. P., Bruulsema T. W., Gehl R. J., Kanter D., Mauzerall D. L., Rotz C. A. and Williams C. O. (2013). Nitrogen-climate interactions in US agriculture. *Biogeochemistry*, **114**, 41–70. https://link.springer.com/article/10.1007/s10533-012-9802-4 (accessed 8 August 2020).

Roosevelt F. D. (1937). Letter to all State Governors on a Uniform Soil Conservation Law. https://www.presidency.ucsb.edu/documents/letter-all-state-governors-uniform-soil-conservation-law (accessed 18 August 2020).

Rosen T. and Christianson L. 2017. Performance of Denitrifying Bioreactors at Reducing Agricultural Nitrogen Pollution in a Humid Subtropical Coastal Plain Climate. Water. **9**(2), 112. https://www.mdpi.com/2073-4441/9/2/112/htm (accessed 17 October 2020).

Rossi M. W., Whipple K. X. and Vivoni E. R. (2015). Precipitation and evapotranspiration controls on daily runoff variability in the contiguous United States and Puerto Rico. *Journal of Geophysical Research: Earth Surface*, **121**(1), 128–145. https://doi.org/10.1002/2015JF003446 accessed online at agupubs.onlinelibrary.wiley.com (accessed 20 August 2020).

Royer T. V., David M. B. and Gentry L. E. (2006). Timing of riverine export of nitrate and phosphorus from agricultural watersheds in Illinois: implications for reducing nutrient loading to the Mississippi River. *Environmental Science & Technology*, **40**(13), 4126–4131. https://pubs.acs.org/action/showCitFormats?doi=10.1021%2Fes052573n&href=/doi/10.1021%2Fes052573n (accessed 12 August 2020).

RssWeather.com. (2020). Climate for Des Moines, Iowa. http://www.rssweather.com/climate/Iowa/Des%20Moines/ (accessed 23 August 2020).

Ruddiman W. (2018). Preindustrial Anthropogenic CO_2 Emissions: How Large? http://www.realclimate.org/index.php/archives/2018/10/pre-industrial-anthropogenic-co2-emissions-how-large/ (accessed 16 August 2020).

Santi C., Bogusz D. and Franche C. (2013). Biological nitrogen fixation in non-legume plants. *Annual of Botany*, **111**(5), 743–767. https://www.ncbi.nlm.nih.gov/pmc/articles/PMC3631332/ (accessed 10 August 2020).

Satterlund D. R. (1972). Wildland Watershed Management. The Ronald Press Company, New York, USA.

Sawyer J. (2014). Estimating Nitrogen Losses in Wet Corn Fields. https://crops.extension.iastate.edu/cropnews/2014/06/estimating-nitrogen-losses-wet-corn-fields (accessed 16 August 2020).

Sazonova T. S., Romanovsky V. E., Walsh J. E. and Sergueev D. O. (2004). Permafrost dynamics in the 20th and 21st centuries along the East Siberian Transect. *Journal of Geophysical Research*, **109**, 1–20. https://agupubs.onlinelibrary.wiley.com/doi/pdf/10.1029/2003JD003680 (accessed 16 August 2020).

Schilling K. E. and Helmers M. (2008). Effects of subsurface drainage tiles on streamflow in Iowa agricultural watersheds: exploratory hydrograph analysis. *Hydrological Processes*, **22**(23) 4497–4506. https://www.researchgate.net/publication/229737999_Effects_of_Subsurface_Drainage_Tiles_on_Streamflow_in_Iowa_Agricultural_Watersheds_Exploratory_Hydrograph_Analysis (accessed 12 August 2020).

Schuur E., McGuire A. G. K. and Romanovsky V. (2018). Arctic and boral carbon. In: Second State of the Carbon Cycle Report (SOCCR2): A Sustained Assessment Report. N. Cavallaro, G. Shrestha, R. Birdsey, M. A. Mayes, R. G. Najjar, S. C. Reed, P. Romero-Lankao and Z. Zhur (eds.), U.S. Global Change Research Program, Washington, D.C., USA, pp. 428–468. https://carbon2018.globalchange.gov/chapter/11/ (accessed 28 August 2020).

Schwartz A. W. (2006). Phosphorus in prebiotic chemistry. *Philosophical Transactions of the Royal Society*, **361**(1474), 1743–1749. https://www.ncbi.nlm.nih.gov/pmc/articles/PMC1664685/ (accessed 26 August 2020).

Science News. (2015). Phosphorus-Rich Dust from Sahara Desert Keeps Amazon Soils Fertile. http://www.sci-news.com/othersciences/geophysics/science-phosphorus-rich-dust-sahara-desert-amazon-soils-02533.html (accessed 9 August 2020).

Sharratt B. S., Lindstrom M. J., Benoit G. R., Young R. A. and Wilts A. (2000). Runoff and soil erosion during Spring Thaw in the Northern U.S. Corn Belt. *Journal of Soil and Water Conservation*, **55**(4), 487–494. https://www.jswconline.org/content/55/4/487.short (accessed 12 August 2020).

Sheng H. and Cai T. (2019). Influence of rainfall on canopy interception in mixed broad-leaved – Korean Pine Forest in Xiaoxing'an Mountains, Northeastern China. *Forests*, **10**(248), 1–11. https://www.researchgate.net/publication/331678507_Influence_of_Rainfall_on_Canopy_Interception_in_Mixed_Broad-Leaved-Korean_Pine_Forest_in_Xiaoxing%27an_Mountains_Northeastern_China (accessed 11 August 2020).

Sherburne M. (2020). Carbon Emissions from Permafrost Soils Underestimated by 14%. https://phys.org/news/2020-06-carbon-emission-permafrost-soils-underestimated.html, on (accessed 8 August 2020).

Sinclair T. R., Holbrook N. M. and Zwieniecki M. A. (2005). Daily transpiration rates of woody species on drying soil. *Tree Physiology*, **25**(11), 1469–1472. https://doi.org/10.1093/treephys/25.11.1469 (accessed 12 August 2020).

Singh K. P. (1996). Managed Flood Storage Option for Selected Levees along the Lower Illinois River for Enhancing Flood Protection, Agriculture, Wetlands, and Recreation

First Report: Stage and Flood Frequencies and the Mississippi Backwater. Contract Report 590, Office of Water Resources Management, Illinois State Water Survey Hydrology Division, Champaign, IL, USA. https://www.ideals.illinois.edu/bitstream/handle/2142/94255/ISWSCR-590.pdf?sequence=1 (accessed 12 August 2020).

Skidmore E. L. and Layton J. B. (1992). Dry-soil aggregate stability as influenced by selected soil properties. *Soil Science Society of America Journal*, **56**(2), 557–561. https://acsess.onlinelibrary.wiley.com/doi/abs/10.2136/sssaj1992.03615995005600020034x (accessed 10 August 2020).

Sleeter B. M., Loveland T., Domke G., Herold N., Wickham J. and Wood N. (2018). Land cover and land-use change. In: impacts, risks, and adaptation in the United States. In: Fourth National Climate Assessment, Volume II. D. R. Reidmiller, C. W. Avery, D. R. Easterling, K. E. Kunkel, K. L. M. Lewis, T. K. Maycock and B. C. Stewart (eds.), U.S. Global Change Research Program, Washington, D.C., USA, pp. 202–231.

Smil V. (2001). Enriching the Earth: Fritz Haber, Carl Bosch, and the Transformation of World Food Production. MIT Press, Cambridge, MA, USA.

Smil V. (2012). Nitrogen Cycle and World Food Production. *World Agriculture, #1203*. www.world-agriculture.net/files/pdf/nitrogen-cycle-and-world-food-production-world-agriculture.pdf (accessed 4 February 2021).

Smith S. V. (1984). Phosphorus versus nitrogen in the marine environment. *Limnology and Oceanography*, **29**(6), 1149–1160. https://aslopubs.onlinelibrary.wiley.com/doi/pdf/10.4319/lo.1984.29.6.1149 (accessed 25 August 2020).

Smith J. (2020). Lake Erie Turns Toxic Every Summer. Officials Aren't Cracking Down on the Source. https://publicintegrity.org/environment/growing-food-sowing-trouble/lake-erie-toxic-algae-farm-manure-runoff/ (accessed 31 August 2020).

Smith J., Hawes C., Karlovits G. and Eberts III R. (2013). CRP Flood Damage Reduction Benefits to Downstream Urban Areas, USACE, Rock Island District Pilot Project Report, Farm Service Agency, U.S. Department of Agriculture, Washington, D.C., USA. https://www.fsa.usda.gov/Assets/USDA-FSA-Public/usdafiles/EPAS/PDF/fsa_pilot_report_final.pdf (accessed 12 August 2020).

Soil Science Society of American. (2013). Soils Matter, Get the Scoop! https://soilsmatter.wordpress.com/2013/08/29/soil-formation/ (accessed 21 August 2020).

Sojka R. E., Bjorneberg D. L. and Entry J. A. (2002). Irrigation: an historical perspective. In: Encyclopedia of Soil Science. R. Lal (ed.), 1st edn, Marcel Dekker, Inc., New York, USA, pp. 745–749.

Stichler C. and McFarland M. 2020. Crop Nutrient Needs in South and Southwest Texas. Agrilife Texas Agricultural Extension https://agrilifeextension.tamu.edu/library/farming/crop-nutrient-needs-in-south-and-southwest-texas/ (accessed 1 November 2020).

Sullivan B. K. (2020). 'Rivers in the Sky' causing billions in flood damage. *U.S. Insurance Journal*. https://www.insurancejournal.com/news/national/2020/02/12/558222.htm (accessed 10 August 2020).

Suszkiw J. (2017). Protecting the Earth's 'Thin Skin'. https://www.usda.gov/media/blog/2017/12/12/protecting-earths-thin-skin (accessed 21 August 2020).

Suszkiw J. (2019). Study Clarifies U.S. Beef's Resource Use and Greenhouse Gas Emissions. https://www.ars.usda.gov/news-events/news/research-news/2019/study-clarifies-us-beefs-resource-use-and-greenhouse-gas-emissions/ (accessed 26 June 2020).

Syvitski J. P. M. and Kettner A. (2011). Sediment Flux and the Anthropocene. https://royalsocietypublishing.org/doi/10.1098/rsta.2010.0329 (accessed 8 August 2020).

Tang C. M. (2020). Jurassic Period. https://www.britannica.com/science/Jurassic-Period (accessed 8 August 2020).

Tanner T. ed. (2012). Aldo Leopold: The Man and His Legacy. Soil and Water Conservation Society, Ankeny, IA, USA.

Telles T. S., Guimarães M. and Dechen S. C. F. (2011). The costs of soil erosion. *Revista Brasileira de Ciência do Solo*, **35**(2), 287–298. https://www.scielo.br/scielo.php?pid=S0100-06832011000200001&script=sci_arttext&tlng=en (accessed 31 August 2020).

Thorman R. E., Nicholson F. A., Topp C. F. E., Bell M. J., Cardena L. M., Chadwick D. R., Cloy J. M., Misselbrook T. H., Rees R. M., Watson C. J. and Williams J. R. (2020). Towards Country-Specific Nitrous Oxide Emission Factors for Manures Applied to Arable and Grassland Soils in the UK, Waste Management. https://www.frontiersin.org/articles/10.3389/fsufs.2020.00062/full (accessed 30 July 2020).

Tilley D. and Winger M. (2014). ID Corn/Cover Crop Interseeding Trial. 2014 Study Report. Aberdeen Plant Materials Center, Aberdeen, ID, USA. https://www.nrcs.usda.gov/Internet/FSE_PLANTMATERIALS/publications/idpmcsr12526.pdf (accessed 13 August 2020).

Tipping E., Somerville C. J. and Luster J. (2016). The C:N:P stoichiometry of soil organic matter. *Biochemistry*, **130**, 117–131. https://doi.org/10.1007/s10533-016-0247-z (accessed 8 August 2020).

Tuttle L. E. (2013). Enhancement of the Floodplain Management Program in Raleigh, NC: An Analysis of Flood Risk Reduction and Preparedness Strategies. Master of Science Thesis, The University of Texas at Austin, Austin, TX, USA.

Ufoegbune G. C., Ogunyemi O., Eruola A. O. and Awomeso J. A. (2010). Variation of interception loss with different plant species at the University of Agriculture, Abeokuta, Nigeria. *African Journal of Environmental Science and Technology*, **4**(12), 831–844. file:///C:/Users/skiph/AppData/Local/Temp/71358-Article%20Text-154123-1-10-20111027.pdf (accessed 11 August 2020).

UNESCO. (2019a). World Water Assessment Programme: Facts and Figures. http://www.unesco.org/new/en/natural-sciences/environment/water/wwap/facts-and-figures/all-facts-wwdr3/fact-24-irrigated-land/ (accessed 11 August 2020).

UNESCO. (2019b). World Water Assessment Programme: Facts and Figures. Food and Agriculture. http://www.unesco.org/new/en/natural-sciences/environment/water/wwap/facts-and-figures/food-and-agriculture/ (accessed 11 August 2020).

United Nations. (1999). The World at Six Billion. Population Division, Department of Economic and Social Affairs, United Nations, New York, USA. https://search.archives.un.org/uploads/r/united-nations-archives/4/b/8/4b8e1e4538ee528ec43c72fcbdd3bddb9718410f3e414240617f97ac98d388e2/S-1092-0117-01-00002.pdf (accessed 19 August 2020).

University of California-Davis. (2020). http://lawr.ucdavis.edu/classes/ssc219/biogeo/lateral.htm (accessed 8 August 2020).

U.S. Agricultural Research Service. (2020). The Problem of Wind Erosion. https://infosys.ars.usda.gov/WindErosion/problem/problem.shtml (accessed 12 June 2020).

U.S. Census Bureau. (2020). U.S. and World Population Clock. https://www.census.gov/popclock/ (accessed 19 August 2020).

U.S. Environmental Protection Agency. (1976). Final Environmental Impact Statement: Crabtree Creek, Wake County, North Carolina Interception Sewer EPA Project C370344, 555.

U.S. Environmental Protection Agency. (2009). Nitrogen Backgrounder: Overview. https://yosemite.epa.gov/sab%5CSABPRODUCT.NSF/B1883A682B97F017852575B6004173A4/$File/Microsoft+PowerPoint+-+Module+1A_022808[donnam].pdf (accessed 31 July 2020).

U.S. Environmental Protection Agency. (2010). 'Methane and Nitrous Oxide Emissions from Natural Sources'. Report EPA 430-R-10-001, Office of Atmospheric Programs, U.S. Environmental Protection Agency, Washington, D.C., USA. https://nepis.epa.gov/Exe/ZyPDF.cgi/P100717T.PDF?Dockey=P100717T.PDF (accessed 22 August 2020).

U.S. Environmental Protection Agency. (2011). Reactive Nitrogen in the United States: An Analysis of Inputs, Flows, Consequences, and Management Options: A Report of the EPA Science Advisory Board, EPA-SAB-11-013, EPA Science Advisory Board, U.S. Environmental Protection Agency, Washington, D.C., USA, https://nepis.epa.gov/Exe/ZyNET.exe/P100DD0 K.TXT?ZyActionD=ZyDocument&Client=EPA&Index=2011+Thru+2015&Docs=&Query=&Time=&EndTime=&SearchMethod=1&TocRestrict=n&Toc=&TocEntry=&QField=&QFieldYear=&QFieldMonth=&QFieldDay=&IntQFieldOp=0&ExtQFieldOp=0&XmlQuery=&File=D%3A%5Czyfiles%5CIndex%20Data%5C11thru15%5CTxt%5C00000003%5CP100DD0 K.txt&User=ANONYMOUS&Password=anonymous&SortMethod=h%7C-&MaximumDocuments=1&FuzzyDegree=0&ImageQuality=r75g8/r75g8/x150y150g16/i425&Display=hpfr&DefSeekPage=x&SearchBack=ZyActionL&Back=ZyActionS&BackDesc=Results%20page&MaximumPages=1&ZyEntry=1&SeekPage=x&ZyPURL (accessed 26 June 2020).

U.S. Environmental Protection Agency. (2015). The Importance of Clean Air to Clean Water in the Chesapeake Bay. https://www.epa.gov/sites/production/files/2015-06/documents/cb_airwater_fact_sheet_jan2015.pdf (accessed 13 August 2020).

U.S. Environmental Protection Agency. (2016). National Rivers and Streams Assessment 2008–2009: A Collaborative Survey (EPA/841/R-16/007). Office of Water and Office of Research and Development, U.S. Environmental Protection Agency, Washington, D.C., USA. https://www.epa.gov/sites/production/files/2016-03/documents/nrsa_0809_march_2_final.pdf (accessed 21 August 2020).

U.S. Environmental Protection Agency. (2020a). Report on the Environment: Water. https://www.epa.gov/report-environment/water (accessed 21 August 2020).

U.S. Environmental Protection Agency. (2020b). Understanding-Science-Ocean-and-Coastal-Acidification. https://www.epa.gov/ocean-acidification/understanding-science-ocean-and-coastal-acidification (accessed 8 August 2020).

U.S. Environmental Protection Agency Science Advisory Board. (2007). Hypoxia in the Northern Gulf of Mexico: An Update by the EPA Science Advisory Board. EPA-SAB-08-003. U.S. Environmental Protection Agency, Washington DC, USA. https://yosemite.epa.gov/sab/sabproduct.nsf/C3D2F27094E03F90852573B800601D93/$File/EPA-SAB-08-003complete.unsigned.pdf (accessed 18 October 2020).

U.S. Global Change Research Program. (2014). National Climate Assessment. https://www.globalchange.gov/nca3-downloads-materials, (accessed online 8 August 2020).

U.S. Global Change Research Program. (2018a). Impacts, risks, and adaptation in the United States: Fourth National Climate Assessment. In: U.S. Global Change Research Program. D. R. Reidmiller, C. W. Avery, D. R. Easterling, K. E. Kunkel, K. L. M. Lewis, T. K. Maycock and B. C. Stewart (eds.), National Oceanic and Atmospheric Administration, U.S. Department of Commerce, Washington, D.C., USA, Vol. **II**. https://nca2018.globalchange.gov/downloads/ (accessed 12 August 2020).

U.S. Global Change Research Program. (2018b). Fourth National Climate Assessment. National Oceanic and Atmospheric Administration, U.S. Department of Commerce, Washington, D.C., USA https://nca2018.globalchange.gov/ (accessed 31 May 2020).

U.S. Grains Council. (2020). Conversion Factors. https://grains.org/markets-tools-data/tools/converting-grain-units/ (accessed 20 August 2020).

U S Geological Survey. (2019). Phosphate rock. In: Mineral Commodity Summaries, National Minerals Information Center. S. M. Jasinski (ed.), U.S. Geological Survey, U.S. Department of Interior. Washington, D.C., USA. https://prd-wret.s3-us-west-2.amazonaws.com/assets/palladium/production/atoms/files/mcs-2019-phosp.pdf (accessed 20 August 2020).

Van Horn H. H., Wilkie A. C., Powers W. J. and Nordstedt R. A. 1994. Components of Dairy Manure Management Systems. *Journal of Dairy Science*, **77**, 2008–2030. https://reader.elsevier.com/reader/sd/pii/S0022030294771472?token=E03CE291926CF3A92C11A B3DE5E59DBCC18600883F79C6DE92197B85F316C1A5F87D003B298EEAEA878D 9946313F205C (accessed 1 November 2020).

Vitousek P. M., Menge D. N., Reed S. C. and Cleveland C. C. (2013). Biological nitrogen fixation: rates, patterns and ecological controls in terrestrial ecosystems. *Philosophical Transaction of the Royal Society*, **368**(1621), 20130119. https://royalsocietypublishing.org/doi/full/10.1098/rstb.2013.0119 (accessed 20 August 2020).

Wallace C. W., McCarty G., Lee S., Brooks R. P., Veith T. L., Kleinman P. J. A. and Sadeghi A. M. (2018). Evaluating concentrated flowpaths in Riparian forest buffer contributing areas using LiDAR Imagery and topographic metrics. *Remote Sensing*, **10**(4), 614. https://www.mdpi.com/2072-4292/10/4/614/htm (accessed 22 August 2020).

Wigley T. M. L. (1983). The pre-industrial carbon dioxide level. *Climatic Change*, **5**, 315–320. https://doi.org/10.1007/BF02423528 (accessed 8 August 2020).

Wikipedia. (2020). Feed Conversion Ratio. https://en.wikipedia.org/wiki/Feed_conversion_ratio (accessed 20 August 2020).

Williams J. T. (1995). The Role of Fire in Ecosystem Management. Gen. Tech. Rep. PSW-GTR-158. U.S. Forest Service, U.S. Department of Agriculture, Washington, D.C., USA. https://www.fs.fed.us/psw/publications/documents/psw_gtr158/psw_gtr158_07_williams.pdf (accessed 11 August 2020).

Williams M. R., King K. W. and Fausey N. R. (2015). Drainage water management effects on tile discharge and water quality. *Agricultural Water Management*, **148**, 43–51. https://www.researchgate.net/publication/266561993_Drainage_water_management_effects_on_tile_discharge_and_water_quality (accessed 12 August 2020).

Wise S. (2013). The Periodic Table. https://starstruckworld.wordpress.com/2013/05/26/the-periodic-table/ (accessed 8 August 2020).

Wolfram Research, Inc. (2020). Abundance in the Universe of the Elements. https://periodictable.com/Properties/A/UniverseAbundance.html (accessed 8 August 2020).

Woli K. P., David M. B., Cooke R. A., McIsaac G. F. and Mitchell C. A. (2010). Nitrogen balance in and export from agricultural fields associated with controlled drainage systems and denitrifying bioreactors. *Ecological Engineering*, **36**(11), 1558–1566. https://www.sciencedirect.com/science/article/abs/pii/S0925857410001102 (accessed 13 August 2020).

Wong J. (2020). Can Higher CO_2 Levels Boost Plant Life Enough to Dent Global Warming? https://www.newscientist.com/article/mg24632820-700-can-higher-co2-levels-boost-plant-life-enough-to-dent-global-warming/ (accessed 9 August 2020).

World Health Organization. (2019). Drinking Water. https://www.who.int/news-room/fact-sheets/detail/drinking-water (accessed 22 August 2020).

World Meteorological Organization, 2010. Greenhouse Gases Reach Record Levels – WMO Highlights Concerns about Global Warming and Methane. World Meteorological Organization Press Release No. 903. C. Richard-Van Maele, Ed. https://public.wmo.int/en/media/press-release/no-903-greenhouse-gases-reach-record-levels-wmo-highlights-concerns-about-global (accessed 3 September 2020).

Worldometer. (2020). World Population by Year. https://www.worldometers.info/world-population/world-population-by-year/ (accessed 19 August 2020).

World Population Review. (2020). https://worldpopulationreview.com/us-cities/raleigh-nc-population (accessed 2 July 2020).

Wortmann C. S., Ferguson R. B., Hergert G. W., Shapiro C. A. and Shaver T. M. (2013). Nutrient Management Suggestions for Grain Sorghum. University of Nebraska–Lincoln Extension. NebGuide G1669, Lincoln, NE, USA. https://agronomy.unl.edu/faculty/ferguson/g1669.pdf (accessed 23 August 2020).

Xiao Q., Hu Z., Fu C., Bian H., Lee X., Chen S. and Shang D. (2019). Surface nitrous oxide concentrations and fluxes from water bodies of the agricultural watershed in Eastern China. *Environmental Pollution*, **251**, 185–192. https://www.sciencedirect.com/science/article/abs/pii/S0269749118353107#! (accessed 22 August 2020).

Zaimes G., Nichols M. and Green D. 2006. Characterization of riparian areas. In: Understanding Arizona's Riparian Areas. G. Zaimes (ed.), University of Arizona Extension Publication. https://extension.arizona.edu/sites/extension.arizona.edu/files/pubs/az1432.pdf (accessed 18 October 2020).

Zehr J. and Ward B. B. (2002). Nitrogen cycling in the ocean: new perspectives on processes and paradigms. *Applied and Experimental Microbiology*, **68**(3), 1015–1034. https://aem.asm.org/content/68/3/1015 (accessed 31 August 2020).

Zulauf C. and Brown B. (2019). 'Use of Tile, 2017 US Census of Agriculture'. *Farmdoc Daily*, **9**, 141.

Zuzel J. F., Allmaras R. R. and Greenwalt R. (1982). Runoff and soil erosion on frozen soils in northeastern Oregon. *Journal of Soil and Water Conservation*, **37**(6), 351–354. https://www.jswconline.org/content/37/6/351.short (accessed 12 August 2020).

IWA Publishing's authorised EU representative for General Product
Safety Regulations is Diane D'Arras, 15 rue Duret, 75116 Paris,
France, e-mail: safety@iwap.co.uk.

Printed and bound by CPI Group (UK) Ltd, Croydon, CR0 4YY
16/06/2026
02140358-0001